HALLEY'S COMET & the PRINCIPIA

History's ancient annals fix
The year one thousand sixty-six
(Then a fiery comet whirled,
Dreadful omen round the world.)
As the time when England's lord
Fell before the Norman sword.

ANONYMOUS

HALLEY'S COMET

Tercentenary Edition

eter Lancaster Brown
& the PRINCIPIA
ith 180 illustrations
aries
press

This tercentenary edition first published 1986 by
Aries Press (of Aldeburgh),
10A St Peter's Road,
Aldeburgh, Suffolk, IP15 5BG, UK

British Library Cataloguing Publication Data
Lancaster Brown, Peter, 1927–
Halley's comet and the Principia.
1. Halley's comet
2. the Principia
I. Title
523.6'4 QB723.H2

ISBN 0-948863-01-3 Hardback
ISBN 0-948863-00-5 Softback

Printed in Great Britain

CONTENTS

... Could a historiographer drive on his history as a muleteer drives on his mule,——straight forward;—— for instance, from Rome all the way to Loretto, without ever once turning his head aside either to the right hand or to the left,——he might venture to foretell you to an hour when he should get to his journey's end:——but the thing is, morally speaking, impossible: For, if he is a man of the least spirit he will have fifty deviations from a straight line to make with this or that party as he goes along, which he can no ways avoid. He will have views and prospects to himself perpetually soliciting his eye, which he can no more help standing still to look at than he can fly; he will moreover have various

Accounts to reconcile:
Anecdotes to pick up:
Inscriptions to make out:
Stories to weave in:
Traditions to sift:
Personages to call upon:
Panegyrics to paste up at this door;

... All which both the man and his mule are quite exempt from. To sum up all; there are archives at every stage to be looked into, and rolls, records, documents, and endless genealogies, which justice ever and anon calls him back to stay the reading of:——In short, there is no end of it...

LAURENCE STERNE
Tristram Shandy

Every thing that relates to great men, even the smallest particulars of their lives, have a charm that renders them peculiarly interesting. All that tends to make us acquainted with their manners, their character, or their genius, is received with avidity.

Publisher JOHN MURRAY (1745-93)

... It is true I lay down a scheme, but fancy is so fertile I often start fresh hints, and cannot but pursue them.

DANIEL DEFOE

The Elements of
CHAP. XVI.
PHILOSOPHIÆ NATURALIS PRINCIPIA MATHEMATICA.
Autore JS. NEWTON, Trin. Coll. Cantab. Soc. Matheseos Professore Lucasiano, & Societatis Regalis Sodali.
IMPRIMATUR. S. PEPYS, Reg. Soc. PRÆSES. Julii 5. 1686.
LONDINI, Jussu Societatis Regiæ ac Typis Josephi Streater. Prostat apud plures Bibliopolas. Anno MDCLXXXVII.
A SYNOPSIS OF THE Astronomy of COMETS.
By EDMUND HALLEY L.L.D. Savilian Professor of Geometry at Oxford.
THE ELEMENTS OF Sir ISAAC NEWTON's PHILOSOPHY.
By Mr. VOLTAIRE.
Translated from the FRENCH.
Revised and Corrected By JOHN HANNA, M. A. Teacher of the Mathematicks.
With Explication of some Words in Alphabetical Order.
Pulchre sibi disputare videntur, cum quod evertere non possunt, tanquam ridiculum contemnunt. Horroccius.
LONDON: Printed for STEPHEN AUSTEN at the Angel and Bible in St. Paul's Church-Yard.
THE GREAT AMATEUR.
38. To NEWTON. 1695.
Honoured Sr.
In answer to yours of the 1st October, I give you many th
The Cata
Year of Discovery
Discoverer
Designation
Perihelion
(Chinese)
(Aristotle)
(Chinese)
(Chinese) ?

To all comet lovers
past, present and future
but in particular to those in 2061...

Author's note

I make no apology for including only half-tone black and white pictures as these are closest to a comet's true colour. On rare occasions, when large comets pass close to the Sun, they may (briefly) take on a yellowish or more rarely a greenish overtone (*see* text), but comets have absolutely no resemblance to those garish multi-coloured illustrations beloved by some modern publishers, authors and imaginative artists.

PETER LANCASTER BROWN

PROLOGUE

In 1682 a young amateur astronomer, just newly married, set up his private observatory in Islington, then a sleepy country village on the outskirts of a rapidly sprawling post-Restoration London. Not long after, at the back end of August, and at about the same time as his friend William Penn was creating his own niche in history by sailing to North America to establish his new "Pensilvania" colony, a bright comet appeared. As may be supposed, this celestial event attracted the young astronomer's whole attention and doubtless kept him from his bride's bed while he diligently tracked it across the backcloth of stars.

The young astronomer, named Edmond Halley (1656-1742), had not been the first to spot the presence of the new comet—this was done by two Frenchmen, Auguste Picard and Philippe De La Hire, at the Paris Observatory some days earlier—so any contemporary fame for its discovery was not his (*see* note 1). With this in mind Halley himself could not then possibly foresee how this same comet would one day bring him a greater, more lasting fame than ever it did to its two discoverers and that long after he was dead his name would be more widely known than it was in his own day and the comet itself would receive world-wide publicity.

Any comet-fame due to Halley then was not forthcoming in his own lifetime. In the seventeenth and eighteenth centuries his reputation as a man of science rested chiefly on other accomplishments now largely forgotten. However, the story of Edmond Halley and his comet is inseparably linked to the story of the unique creation of his friend Isaac Newton's great scientific work *Principia*, published in 1687. Without the *Principia* Halley would not later have been able to predict the return of the bright comet he had seen in 1682; equally important, without Halley it is doubtful if the *Principia* would ever have been written...

Among transient celestial phenomena, comets are perhaps the most intriguing of all because of their often unexpected and sometimes spectacular appearances in our skies. Before the time of Newton and Halley very little was known about them. All the ancient civilizations were equally puzzled by them. Their presence, however, was generally thought to have profound astrological significances, especially in their role as harbingers of evil—either to signal the imminent death of a notable or an outbreak of plague, war or famine.

Until recently, the name comet was synonymous with prodigy, presage or portent, and ordinary people once used it to mean punish or scourge. It is no coincidence either that the word disaster, from Latin *dis* and *astrum* (star destiny) is a direct referral to the traditional role of comets while influenza is a word coined (from the Latin *influentia*) for its allusion to comets as carriers of disease.

The ancients were unsure whether comets were true astronomical bodies or whether—owing to the weighty contrary opinion of Aristotle in the matter—they were phenomena occurring locally in the Earth's atmosphere. In the first century AD the ill-fated Seneca, tutor to Nero, was quite sure Aristotle was mistaken yet neither he nor anyone else could prove it either way. Seneca, however, was optimistic about man's rapid progress in scientific ideas and predicted that one day a scholar would arise who would prove beyond doubt that comets were wanderers and akin to the planets. Little did Seneca realize that it would take his successors 1,600 years after his own time to reach even a half-truth about comets and to arrive at the *frontiers* of the full truth another 300 years would pass by.

Today, in spite of the rapid advances of knowledge made in the space age, we are still unsure about the true nature of comets and how they originate. We know that in the past comets have brushed with the Earth and some have collided with it. Several times in its long history the Earth has passed through the tail of Halley's comet as it did as recently as 1910. What long-term effects this may have on the evolution of life on our planet is perhaps the most controversial question relating to comets. Some contemporary astronomers, like the astrologers of yesteryear, believe comets carry organic matter and pathogens spawned in remote space. These are then shed in their tails and eventually penetrate our atmosphere—later to become incorporated into living cells. While these ideas would have seemed downright absurd to scientists just a few years back, the question of the seeding of life from space now requires serious consideration. Halley's comet in 1986 provided the first close-up view of a comet nucleus and some tantalizing evidence to support the life-from-space ideas (Chapter 20). Whether comets like Halley's are the source of genesis on Earth is still conjectural. However, what is clear is that Halley's comet holds a particular fascination for mankind rarely equalled by any other celestial body.

1 Halley's comet photographed from Australia on 22 February 1986 showing a disruption in part of the gas tail and a spike extended towards the Sun.

CHAPTER 1

Prodigies, Plague and Fire

"I think we get the best view of any subject if we look at the natural growth of things from the beginning."—ARISTOTLE

Eighteen years before Halley observed the comet that one day would immortalize his name, another bright comet had been widely observed. It was discovered in Spain on 17 November 1664 and during the following days it was seen all over Europe. It appeared over London when war with the Dutch hung in the balance and round the time a 31-year-old, up-and-coming navy clerk by the name of Samuel Pepys was agonizing over the prospects for further dalliance with the wife of a ship's carpenter.

Although Pepys claimed to have seen the new comet on 24 November, this was only an idle boast to impress friends, and it was not until weeks later he caught his first glimpse of it. It was, however, seen early on by his friend and fellow-diarist John Evelyn from his house at Sayes Court in Deptford and by many others in London. One almost certainly, among tens of thousands of other curious Londoners, was 8-year-old Edmond Halley, viewing from the garden or attic window of his father's town house in Winchester Street just across the City from Seething Lane. It would be young Edmond's first ever sight of a comet, and he would have been given news of it from his father's apprentice soap-boiler—an older, highly intelligent, lad who had already taught his master's son the rudiments of astronomy, arithmetic and writing.

Another one to watch it, just a few streets away, was Robert Hooke, the 29-year-old curator of experiments at the recently chartered Royal Society. Hooke was colleague and close friend of Christopher Wren who observed the comet from Oxford where he was Savilian Professor of Astronomy.

At England's other university, Cambridge, north of London, the comet was eagerly noted by an obscure 21-year-old undergraduate called Isaac Newton. Not content to view the comet simply as an entertaining spectacle like his fellow undergraduates, the brilliant, neurotic student sat up every clear night to plot its course across the background stars from his study window in Trinity College.

Pepys then knew nothing of young Halley or Newton, although in the years to follow he came to know both well; but he already knew about Hooke, at least by reputation, as being the leading virtuoso of the Royal Society—or "the cleverest man in England" as Evelyn himself called him. The new comet was the talk of the London taverns and the coffee-houses; coupled with Hooke's name as the foremost, learned observer of it, the subject was on everyone's lips. When Pepys heard that Lord Sandwich, his patron, had also been making precise observations of the comet from his flagship in the Channel, he was now determined not to miss out on the spectacle himself.

Arrangements were made to view it from Tower Hill. On the early morning of Christmas Eve 1664, Pepys, accompanied by his new servant Tom Edwards, ventured forth into the bright moonlit, frosty night. Yet disappointment awaited them. In Pepys's own blunt words: "no Comett to be seen; so after running round the Hill, I and Tom, we home and then to bed."

Pepys was no stargazer and had not even known where to look. However, not to be deprived of the spectacle and on taking advice, he tried again the same night after returning home from his office. "This evening, I being informed, did look and saw the Comett." But Pepys thought it much like a star, only larger and duller than others visible in the sky. It was hardly the image of the comets he had seen illustrated in prints like those of the great comets described by Johannes Kepler back in 1607 and 1618. This new comet showed no tail yet Hooke in London, and Sandwich from Portsmouth, had spoken of a long tail that swept back from the body of the comet. Pepys failed to realize the dimming effect of the powerful light of the Moon. This was partly the reason for his failure to see the comet the previous morning. A tail was there, sure enough, but almost blotted out by moonlight. It still stretched 20° behind the comet when Johannes Hevelius saw it from his observatory in Danzig on 12 January 1665.

To Pepys, with the prospect of war with the Dutch, the comet must have looked threatening. He would have read Lilly's current almanac and the hysterical broadsheets of the other leading astrologers. It was easy to scoff at these men's dire predictions about comet prodigies, but, this time, would they be proven right?

Pepys, as a London citizen, was not the only one to suspect the new comet ominous. Evelyn himself had a lifelong suspicion about the role of all comets. In the new year the eclipse of the Moon in January, while the comet was still visible, only added fuel to those who forecast dire consequences. William Lilly, the one-eyed London astrologer reputed to have three wives and whose outpourings Pepys often read avidly, confidently predicted "the Sword, Famine, Pestilence and Mortality or Plague".

John Gadbury, a well-known comet-astrologer who was reckoned to have a little more science up his sleeve than his competitors, forecast the new prodigy meant the return of the Black Death. There had been five other serious plague attacks: in 1527, 1593, 1603, 1625 and 1636. On these occasions, claimed Gadbury, either a comet or an eclipse, or both, had been seen before the onset of the epidemic. Now from this latest comet he feared even worse evils.

Pepys, like other citizens, was probably reassured when they read those who scoffed at the astrologers. Nevertheless, at the level of the common-street citizen, hysteria prevailed. Puritan churchmen conscious of the unexpected gift from heaven and keen to exploit it, increased the fear when they warned their congregations about Gehenna and the New Babylon and Sodom and Gomorrah. When word got about, this had the desired effect, and people flocked to the City's churches to hear what their preachers had to say about their immediate future and the heavenly significance of the ominous hairy star.

As the weeks passed, Pepys himself was curious to learn more about the comet; he was now only too aware of his own ignorance of science. For some time he had toyed with the idea of joining the ranks of the virtuosi. It was the advent of the comet which persuaded him and on 8 February 1665 he had his friend Thomas Povey propose him and the following week he signed the

2 William Lilly (1602–81), the one-eyed astrologer extraordinary whose dire predictions came true and made the City authorities suspicious he had a direct influence in causing the Great Fire.

3 Comet-astrologer John Gadbury (1627–1704).

Charter-book of the Royal Society as a newly elected Fellow.

War with the Dutch now seemed inevitable, and when hostilities were declared official on 22 February 1665, the ignorant naturally blamed the comet. Had not the astrologers predicted a "Sword-war"!

In spite of pressing business at the Navy Office, and an occasional diverting dalliance, Pepys immersed himself in the weekly activities of the Royal Society whenever he could. He made a special point of attending Hooke's second lecture on the comet when Hooke claimed it was a reappearance of that seen by Kepler in 1618.

Then, on 26 March 1665, yet another bright comet was discovered at Nuremberg and was soon sighted over London. Even those who had scoffed at the significance of the first began to think again. A double presage must have especial significance. The Puritan divines now hotted up their campaigns and warned their flocks of a coming judgement upon London itself for its past sins. Humphrey Smith, a celebrated Quaker, went a step further and published a pamphlet entitled *A Vision* in which he claimed to have seen a burning London...

When in due course the Great Plague broke out in the summer of 1665 and then the Great Fire swept through the City in September 1666, the astrologers had a field-day. First war with the Dutch, then pestilence, followed by fire! When sceptical

4 A very rare medal c 1666 commemorating the London Plague and Great Fire and depicting the two comets believed to be their cause. On one face the comets denote "pestilence, fire and war by which God punishes"; the other face reflects "tranquillity, forgiveness and the blessing of peace and plenty".

scientists of the Royal Society argued that long before the coming of the previous two comets similar calamities had been forecast many times to the point of tedium, the ignorant citizens knew better and turned a deaf ear.

While the advent of the two bright comet prodigies of 1664 and 1665 reinforced the superstitions of the ignorant, the widespread interest it created also heralded the dawn for more scientific studies of them. Two decades later, Halley, Newton, Wren and Hooke, prodigies of another kind—and even the unscientific Pepys himself—would all be closely involved in events leading to the publication of the greatest scientific work ever written. But before that could happen, two, even brighter, comets would intervene to influence all their destinies.

When the first of the bright comets appeared in November 1664, Isaac Newton had already completed three years of intensive study as undergraduate at Trinity College and was on the threshold of a brilliant academic career. Introverted since childhood, he had read much about the sinister influences of comets during his early years spent in rural Lincolnshire, but this was his first opportunity to study a comet at first hand.

He was born on Christmas Day 1642, in the year Galileo died, and raised in the manor-house at Woolsthorpe, a small village south of Grantham. Already fatherless at birth (Isaac senior died three months earlier), the fragile, premature infant was not expected to live. Yet, contrary to all expectations, three years later the once sickly mite had grown into a normal child. It was now he experienced his first trauma in life. His mother, Hannah, remarried, but the new husband, a man named Barnabas Smith, rector of North Witham the village a-mile-and-a-half away, had no intention of taking on the 3-year-old boy. As a consequence young Isaac was left in the care of his maternal grandmother (Ayscough) and remained back at Woolsthorpe.

Being deprived of his mother's affection during these critical years of his life probably had a profound and lasting effect on Newton's personality. One consequence was for neurotic tendencies to develop which were to beset him throughout his youth and adult life. In middle age they would lead him to the verge of mental breakdown.

Meantime, left much to his own devices at Woolsthorpe, Isaac read all the books on hand. He showed an unusual precocity and developed a practical talent for model-making. His nature at this period showed a mischievous bent. After reading about the evil reputation of comets, he designed lanterns of crimped paper, cunningly illuminated them with candles, and then flew them surreptitiously by night from the tails of kites... As anticipated by the young hoaxer, these phantom comets beguiled the local rustics and put the fear of God into them.

5 An eighteenth-century drawing of Newton's birthplace at Woolsthorpe. Today the dormer windows are missing.

It was not, however, until the comet of 1664 appeared in the night skies above Trinity College that Newton had opportunity to renew his interest in the real thing. The comet soon engrossed him. His contemporaries later recalled how the strange undergraduate neglected all else to study it. He even forgot meals, and his cat grew fat from food left uneaten on his master's tray. By now the brilliant student had already slipped into the role of college recluse, shunning all but necessary human contact.

Further confirmation of the early obsession with the comet of 1664 came to light through the husband of Newton's niece, John Conduitt. As a relation by marriage, Conduitt was privy to many of Newton's later-life gossipy reminiscences and tells us: "He [Newton] sate up so often long in the year 1664 to observe a comet.... he found himself much disordered and learned from then on to go to bed betimes."

This anecdote, however, is only a partial truth, for Newton must soon have relapsed into his former bad habits. Throughout his adult life it was common knowledge that once a problem had seized him, and demanded a solution, sleep was a luxury readily dispensed with.

Edmond Halley came from a totally different background to Newton. He was born in Haggerston, on the rural fringes of London, on 29 October 1656. His father, also an Edmond, was at this time a wealthy soap-boiler who owned two homes and much City property. Young Edmond spent his early life between his father's country house and his town house in Winchester Street, near the old City walls. He enjoyed a close and affectionate relationship with his mother, Anne, until her premature death in October 1672. There is every reason to suppose his earlier childhood, in contrast to Newton's, was a secure and relatively happy one. During the height of the Plague the family probably remained in Haggerston. The only known events which marred the family's affairs were the death of his infant sister, Katherine, born in 1658, and the consequences of the Great Fire when much of Edmond senior's City property was razed to the ground. Even so it seems the family remained relatively well off—affluent enough to send young Edmond and his brother, Humphrey (who died abroad in 1684), to a fee-paying public school and support them afterwards as young gentlemen of leisure.

In young Edmond's Pepysian London there were still walls standing thirty feet high. It was through gates set in these same walls Henry V had ridden after his famous victory at Agincourt. Since Agincourt, however, London's population had long since spilled over and spread like a blight into the surrounding countryside.

In the 1660s London housed over half a million—one-tenth of the United Kingdom's population. During Halley's lifetime, under the impetus of the Restoration and the Quiet Revolution, the City's population doubled itself.

Although Halley lived in or near the great metropolis, his early years, in contrast to Newton's well-documented ones, are very shadowy. He was eight years old when the Plague and Fire comets appeared in 1664 and 1665 and only ten when the Fire itself destroyed much of his father's property. The Fire also destroyed the fashionable St Paul's School, lying in the shadow of the great cathedral. The old school dated back to medieval times and among its pupils counted Milton, Marlborough and Pepys. In Pepys's day it was the largest fee-paying school in London; its foundation members were supposedly limited to 153 boys, the number of the fishes caught in the miraculous draught, and most of its pupils were day-boys rather than boarders. It was the natural place where wealthy city merchants aspired to send their sons. The date young Edmond Halley entered its portals is uncertain—owing to the confusion of records as the result of the Fire. Halley's name appears on registers in the late 1660s while the school was temporarily accommodated in premises in Wandsworth, six miles to the south. By his own admission, he was already adept as a scholar before entering St Paul's, and John Aubrey, in his *Brief Lives*, confirms: "At 9 yeares old, his father's apprentice taught [Halley] to write, and arithmetique."

His first High Master was the famous Samuel Cromleholme, who pronounced his name "Crumlum", a man of wide learning reputed to have had the best private library in London until it was destroyed by the Fire. When Cromleholme died, he was replaced as High Master by Dr Thomas Gale who married Barbara Pepys, daughter of Pepys's cousin Roger. By Aubrey's account Halley's stay at St Paul's must have overlapped both high masters. It was probably under Cromleholme Halley was made school captain, which indicates he was talented as an athlete as well as a scholastic and a popular boy with masters and fellow pupils alike. If the curriculum of St Paul's had been weak in arithmetic in Pepys's days, as is known, it was not, apparently, by the time Halley attended twenty years later. Pepys had trouble even with simple multiplications right into his adult life. Halley specialized in geometry and, according to Aubrey, "at 16 could make a dyall, and...thought himself a brave fellow". Aubrey also relates that while Halley was at school "he was very perfect in the caelestiall Globes in so much that I heard Mr Moxton (the Globe-maker) say that if a star were misplaced in the Globe, he would presently find it".

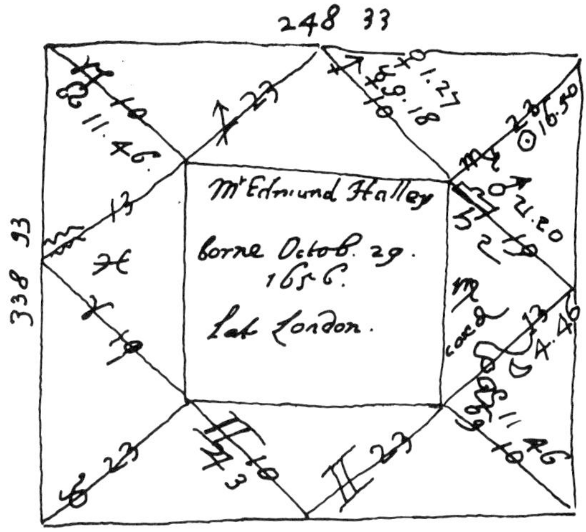

6 A facsimile of Halley's horoscope based on one found in the Bodleian Library, Oxford. The astrologer who cast it, at the instructions of John Aubrey, is not known with certainty, but it was probably either John Gadbury or Henry Coley, the adopted son of William Lilly.

In spite of Halley's interest in arithmetic and astronomy, the school curriculum was the customary and mandatory fare of Latin and Greek, plus Hebrew for top-form boys, just as it had been in Pepys's day. Latin was the lingua franca of European scholarship, and without it no one could hope to succeed either as a gentleman or as a scholar.

It was in the summer of 1673, at the age of seventeen, Halley left St Paul's and entered Queen's College, Oxford as a commoner. When he arrived to take up residence there, he entered the university armed (as one contemporary account has it) with a "curious apparatus..." which included a 24-foot-long telescope and a large quadrant provided by his indulgent father. From the start Halley was determined to make his mark in the world of astronomy.

Twelve years earlier Newton had arrived at Trinity College in very different circumstances. After showing promise at the local grammar school at Grantham, he had become a rebellious teenager living entirely in a world of his own making. By now his stepfather was dead, and Newton was reunited with his mother and his new stepbrother and sisters. As the eldest of the children, his mother wished him to take an interest in the running of the family farm, but he objected strongly to the idea of having to spend the rest of his life shovelling dung. He was truculent to house servants and local rustics alike, and his behaviour was beyond comprehension to all. He was constantly accused of being surly, lazy and rather foolish. In the end his mother finally despaired she could change him and agreed with others he was "fit for nothing, but the 'versity".

Trinity College was then the premier college at Cambridge, admitted by all to be "the famousest College in the University". Newton was entered as a subsizer—a poor student who earned his keep by waiting at table and performing menial tasks for those students who paid for the privilege of dining at high table. A typical day's duty was to rouse other students for morning chapel, act as valet polishing the privilegeds' boots, dressing hair, then wait at table as a flunky—scurrying along corridors and upstairs with orders to the buttery. Another morning duty was emptying the chamber-pots. For someone who had recently protested at shovelling dung this was an ironic turn-about.

Certainly the Newtons were far from poor. The family had the status of local gentry. It must have been hard for Newton, as heir to the lordship of the manor, to swallow his pride and knuckle down to it simply because his mother had refused to support him at university. It is probable that her non-cooperative attitude was in hope he would soon change his mind and come home to help work his heritage. If he was ever tempted to throw in the sponge, we hear nothing about it even in second-hand reminiscences. What we do discover, however, is that in 1662, a year into his undergraduate studies, he underwent a deep religious crisis. The record of this very private experience is written in cipher using the same popular Shelton system of shorthand as employed by Pepys in his diary about the same time. The entries, a recitation of past sins, reveal Newton's neurotic tendencies. By twentieth-century standards some of his past sins appear quite whimsical: "...Making pies on Sunday night... Squirting water [as a boy] on Thy day." Other entries make clear his now serious puritanical streak: "Not loving Thee for Thy self" and "Fearing man above Thee." Several other entries reveal how much he still hated his now dead stepfather for separating him from his mother.

In April 1664 his ordeal as a lowly sizer came to an end when he received a scholarship. It earned him four more years of study and in 1668 his M A degree and a much-welcomed college fellowship.

CHAPTER 2

St Helena, Danzig and the Royal Society

Soon after entering Oxford, Halley began a correspondence with John Flamsteed, created first Astronomer Royal by Charles II in 1675. Flamsteed was ten years older and of a totally different personality to the extrovert Halley. In spite of this, a tentative friendship prospered. Apparently Flamsteed was deeply impressed by the breadth of astronomical knowledge possessed by this unknown student at Oxford. Halley had been fortunate to study under two excellent mathematicians, and his native talent was quickly brought to maturity. Already he was a much superior mathematician than Flamsteed himself. Soon Halley met the Astronomer Royal face to face and was frequently observing with him at Greenwich during the university vacations.

However, as an undergraduate at Oxford, Halley grew restless. Since his incumbency as school captain at St Paul's, there was a man of action as well as an academic craving to find fulfilment. His interests were catholic and he read widely. Some of the books he studied were accounts of voyages and expeditions of fellow Europeans who sought adventure and geographical prizes abroad in the little-charted seas. By 1676 he had been at Oxford for three years, and the same old academic routine bored him; he was now impatient to get on with the business of living.

It was probably the departure of one of his undergraduate friends for the exotic climes of Jamaica that inspired his own impetuous break with Oxford. His intention now was to seek an early fame by mounting his own small expedition to St Helena in the South Atlantic. In the southern hemisphere he could fulfil his ambitions: he could find adventure to satisfy his restless spirit and at the same time he could justify the journey to such a remote spot by carrying out serious scientific work and extend the famous star catalogue of Tycho Brahe to latitudes unseen by the famous Danish nobleman astronomer who had died in 1601.

7 Tycho Brahe (1546–1601) was very influential in the history of cometary astronomy. He proved that the comet of 1577 lay beyond the Moon, and his painstaking observations of Mars provided his protege Kepler with the means to discover the ellipticity of planetary orbits. It was this knowledge which, in Newton's hands, finally led to the understanding of how comets moved.

What reactions his tutors and his father had to this plan is unknown. They likely tried to persuade him to remain at Oxford and first take his degree. After that there was time enough for adventure and his gallivanting. If they did try to persuade him, they failed. Whatever Edmond senior thought of his son's plans, he agreed, in spite of his dwindling fortune, to cover all expenses.

St Helena was not Halley's first choice. He would have preferred the southern tip of Africa, but the Cape was in Dutch hands. Another thought was to go to Rio de Janeiro, but that too was found to be impractical. The enterprise in terms of manpower was a modest one. Through Flamsteed and other influents, who solicited the aid of Charles II, Halley obtained free passages to the lonely British colony via a ship of the East India Company for himself and a man who is mentioned as his friend, a Mr Clerke (or Clark). Who this Mr Clerke actually was is not known. He was possibly and older, paid companion taken along at the insistence of Halley's father, aware that his son was still only twenty years of age. It was also possible the pair were accompanied by Clerke's wife to act as housekeeper during their stay on the island. There is no reference to her in the records of the East India Company, but mention was made of "a woman and her husband" in connection with Halley by Aubrey in his *Brief Lives.*

The account of Halley's stay on St Helena is extremely sketchy. He set up his observatory close to the spot where Napoleon Bonaparte would later find a temporary grave, but he found the island's climate very unsuitable for astronomical observations. We also know he experienced difficulties with the governor, but this was not unique to Halley, and the man was later recalled to London because of unsatisfactory conduct. Less is known of the truth behind the gossip that followed young Halley back to London, but a whisper of gossip there was. It was later to find voice in Aubrey's (often unreliable) *Brief Lives*, which was not to appear in print until 1813 in *Letters by Eminent Persons* , and this relates: "... There went over with [Halley] (amongst others) a woman, and her husband, who had no child in several years, before he came from the Island, she was brought to bed of a child..."

The implication is obvious; the truth uncertain. Aubrey was always capable of outrageous statements about some of his subjects. Nevertheless, Flamsteed, a very pious clergyman, seems to have got wind of it and he made several oblique references to it in later years when his friendship with Halley was strained.

Whether the husband in question was Clerke must remain doubtful. Aboard the East Indiaman, the *Unity* , whose log of the voyage is lost, were likely immigrant passengers who, following the Great Fire and its aftermath, had been encouraged to seek a new life in this southernmost English colony. It is possible the woman and her husband were London immigrants who became acquainted with Halley and Clerke during the long three months' voyage south and whom Halley may have employed as temporary domestics during his stay on the island.

If there were gossip immediately after his return to London in May 1678, it did Halley no harm. By November 1678 the new star catalogue and planisphere of stars was in print and announced at a Royal Society meeting by Robert Hooke, who had also befriended Halley.

In arranging his voyage, Halley had secured the good will and patronage of Charles II. Now, with an eye to his own future, he presented a copy of his new planisphere to the King. As a dabbler in science and patron to the Royal Society, no doubt Charles was delighted to learn he had gained immortality among the southern stars in the form of a new constellation which the opportunist Halley had specially created out of part of the old constellation of Argo, first formed by the ancient Greeks. The new star grouping, named Robur Carolinum, was to commemorate the oak tree in which the King had taken refuge after the Battle of Worcester in 1651.

8 The new constellation Robur Carolinum was formed by Halley in 1679 out of part of Argo Navis in memory of the oak tree in which Charles II hid after the Battle of Worcester in 1651. Here depicted in Hevelius's star atlas of 1690.

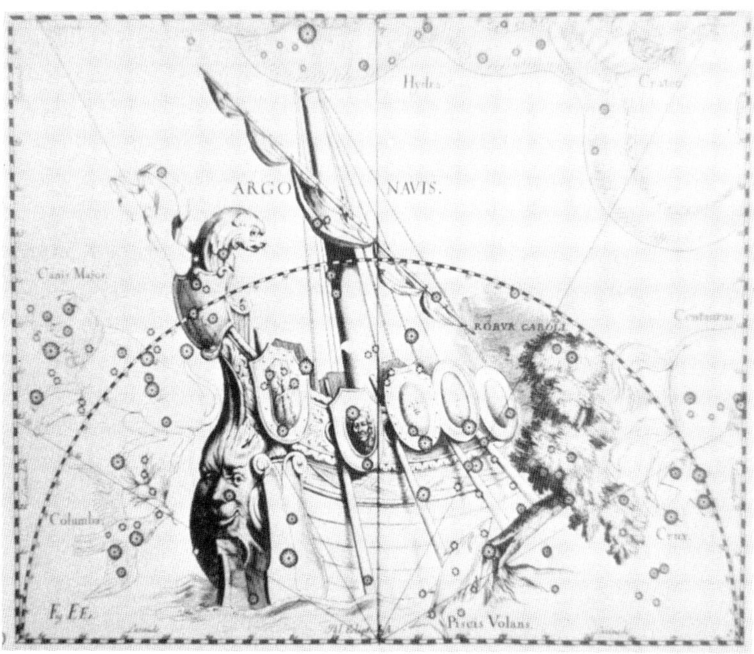

Not surprisingly, Halley's flattering enterprise gained Charles's attentions. For his efforts he was awarded his M A degree at Oxford plus the coveted honour of election as Fellow of the Royal Society.

Halley had accomplished what he had set out to do. He had made his first substantial mark in the world of science and now rubbed shoulders with the establishment as one of its peers. Even the sometimes reticent Flamsteed was much impressed with his protégé's new catalogue of stars and afforded him a personal and very flattering accolade by naming him "the Southern Tycho". It was a compliment the Astronomer Royal was soon to regret.

Halley's star was rising fast. Not trained for commerce and still supported by a generous allowance from his father, he quickly fell into the role of a dilettante virtuoso. By night he busied himself in astronomical observations and during the day in the affairs of the Royal Society. Soon, however, he was on his travels again, this time as the Society's official delegate-ambassador to Johannes Hevelius who had established an observatory in Danzig.

At this time Hevelius was the leading observational astronomer in the world—with a special interest in comets. In 1668 he had published the fruits of his early work in a book entitled *Cometographia* soon to be accepted as the standard work on the subject. Hevelius, however, was equally well known for his long-term work in attempting to establish the most accurate star catalogue by refining the positions of stars observed earlier by Tycho Brahe. His observatory in Danzig was reputed to be the finest in the world and was a mecca for visiting astronomers.

The principal reason for Halley's official visit was to find out at first hand about Hevelius's observing methods and then report back to the Society. It was Hevelius's claim that the now old-fashioned visual sights, much like those Tycho Brahe had used in the pre-telescopic age before 1609, were equal or superior to the recently adopted optical ones. Hooke and Flamsteed, with good reason, hotly disputed this, affirming optical tubes were infinitely superior. The Society, as an international scientific body with whom Hevelius conducted a regular correspondence, wished to maintain cordial relations. Young Halley, it was considered, was just the man to play scientific diplomat. He was

9 Johannes Hevelius (1611–87) portrayed round the time he published his *Cometographia* in 1668.

10 The title-page of Hevelius's *Cometographia* (1668). This book ntains some of the earliest hints about the parabolic nature of comet orbits.

11 Hevelius's open-air observatory in Danzig illustrating a battery of measuring-instruments and one of his long-focus aerial telescopes.

12 Johannes Hevelius, assisted by his wife, using a large double octant in Danzig. It was with instruments of this kind Hevelius claimed he could observe star positions to the same degree of accuracy as those equipped with telescopic sights used by Hooke and Flamsteed.

now an expert himself: he had extended Tycho's work to the southern hemisphere and was someone who, unlike Hooke and Flamsteed, was considered to have no prejudice towards the issue in contention.

In 1662, Hevelius's first wife had died. The following year, when the Danzig astronomer was in his fifties, he had courted and married the beautiful 16-year-old daughter of one of the city's rich merchants. She soon proved to be a willing and invaluable assistant. He taught her how to observe, and prints of the period depict her helping her famous husband to measure star positions with a quadrant (Fig 12).

In 1679, when Halley arrived in the city, Elizabetha Hevelius had matured into one of the city's most handsome creatures. It must have been a remarkable and novel experience for the tall, good-looking and fair-haired Englishman to be briefed in the use of Hevelius's famous long telescopes, one measuring 90 metres, by a highly intelligent member of the opposite sex.

Whether he was tempted ... whether he made overtures ... whether she responded to his winning ways and a young man's courtesies is uncertain. But the facts are that in later years there was a strong whiff of gossip about his alleged liaison with her, even though one of its later tellers, Thomas Hearne, believed it untrue (*see* note 2). On 13 November 1713, on the occasion Halley's portrait was placed by the side of Hevelius's in the Bodleian Library gallery, Hearne was provoked into writing: "And some persons say that he is very justly placed by Hevelius, because he made him (as they give out) a Cuckold, by lying with his wife when he was at Dantzick, the said Hevelius having a very pretty woman to his Wife who had a very great Kindness for Mr Halley, and was (it seems) observed often to be familiar with him. But this Story I am apt to think is false."

Whatever the truth, it seems Hevelius himself was a perfect host to Halley, and the two were soon firm friends. This, however, put Halley in a difficulty, for although he stated he was initially impressed with the accuracy of Hevelius's non-optical observing methods, and duly noted: "so I dare no more doubt of his [Hevelius's] Veracity," it seems Halley was only being diplomatic. In a letter to a friend long after he had returned to London he confessed to having concealed his real opinion of his host's outdated observing technique for fear of bringing on an early departure to the grave of "an old peevish Gentleman". Halley then perhaps, if truth be known, had his own very personal reasons for staying sweet with Hevelius while prolonging his two months' stay in Danzig.

Whatever Halley's true feelings about Hevelius's observing technique and for his wife, he maintained a correspondence with them after his return to London. In his report back to the Society, Halley leaned favourably towards Hevelius. Flamsteed on his part was displeased with the tone of his ex-protégé's report and believed "[Halley] hath been too lavish in his comendation". This and other matters were to open a rift between the pair which was to grow ever wider during Flamsteed's lifetime. In later years, meantime having heard gossip about Danzig and St Helena, the pious Flamsteed was to write to Newton: "...I have done with him [Halley] who has almost ruined himselfe by indiscreet behaviour & you shall hear no more of him from me till wee meet when I shall tell you his history which is too foule and large for a letter." In yet another blast to Newton, Flamsteed also suggested pointedly that "Halley give us a true account of all his St Helena exploits".

Newton had been elected to the Royal Society back in 1672, but after submitting his paper on the theory of colours and clashing with Hooke, who was still demonstrator to the Society, he rarely bothered to travel up to London from Cambridge to attend weekly meetings. Newton's own star had ascended rapidly once his genius was revealed. His brilliance as an innovator and mathematician had been recognized in 1669 when he was elected Lucasian Professor of Mathematics in place of Isaac Barrow, who resigned in favour of his brilliant, old pupil.

During the late 1660s, alchemy had begun to fascinate Newton, and he totally absorbed himself in its literature. But it was his mathematical genius and his remarkable invention of the reflecting telescope, made by his own hands in 1668 and then presented to the Royal Society, which elevated him to the position of leading virtuoso. Even the great Christiaan Huygens in Holland referred to the "marvellous telescope of Mr Newton". Hooke, jealous of his own premier position in British science and not ready to indulge a usurper, claimed that four years previous to Newton's he had invented one himself. This, according to his boast, he had made as a little tube about one inch in length to put in his fob, and it performed more like "a refracting telescope of fifty feet long". If Hooke did invent such a miniature instrument, neither it nor its unique design has survived. James Gregory, however, prior to Newton, had actually published a plan for a reflecting telescope in his *Optica promota* in 1663. The Frenchman Cassegrain, of whom little is known, also claimed to have preceded Newton, but his invention was likely contemporary with Newton's. Either way Newton's optical arrangement was different from the others. It was much more simple and practical, and it was certainly Newton who invented the first reflecting telescope *that worked*.

Since his religious experience back in 1662, Newton had become deeply involved with beliefs in Arianism. As a religious fanatic he turned in on himself and relived the terrible struggles of the Church during the fourth century and came to adopt Athanasius as his personal enemy. By the 1670s he was going through a new crisis in his life and because he felt uncertain about his future income, he wrote to the Royal Society in 1675 requesting it excuse him from payments.

During the 1670s the Royal Society itself was gradually slipping into decline. It had grown from a small group of men with mutual philosophical interests who during the Interregnum had begun to meet at Wadham College, Oxford, to try to discover something of the world they lived in. Underlying this new spirit of enquiry by educated men was the fundamental idea that the movement of everything in the universe was related, and that one had only to understand the divine machinery to discover the elixir of life. Nevertheless, some learned men still held astrological beliefs as the Babylonians had done two thousand years before them. They supposed the movements of the heavenly bodies governed the diseases of the corporal body; that certain days were good for bleeding while others not; that charm and incantation could expel influences from the system.

Following the Restoration, the small Oxford group set about to rid English learning of its mumbo jumbo. Through the efforts of Christopher Wren and others, the patronage of Charles II was obtained and then a royal charter. In this way the Royal Society came about, and a meeting-place in a room in Gresham College, in the very heart of London, was allocated to their activities.

Gresham's in Bishopsgate, just inside the old Roman wall, was itself a remarkable institution. The college building was an Elizabethan-age town house; a two-storey rectangular mansion as large as a castle, and although lying in the very heart of the capital, it had a life and solitude of its own. It took its name from Sir Thomas Gresham, Elizabeth's foremost merchant, who is now best remembered for his dictum, Gresham's law: "bad money drives out good." In his will Sir Thomas stipulated his school should be more practical than the Oxford and Cambridge colleges. He decreed there should be professors of Civil Law, Divinity, Astronomy, Physics, Geometry, Rhetoric and Music; in addition all these subjects must be taught in English, not Latin as at Oxford and Cambridge, and instruction made available to those citizens who could not attend the universities. It was part of the new learning and the Open University of its age and became associated with the name "The Invisible College". Inside its precincts only politics and theology were forbidden as topics for discussion. On all other matters there was encouraged a wide diversity of opinion, for this was held to be the only way to get to the very truth of life.

One of the Gresham astronomy professors was Christopher Wren, who held this position from 1657 to 1661 before accepting the Savilian Professorship at Oxford. Robert Hooke was appointed as Professor of Geometry in 1665. Both these men were founder members of the Royal Society; after the Great Fire in 1666 both worked amicably together to plan and rebuild the devastation of the City.

It has been said that a strong motivation for Charles II's interest in providing a royal charter for the Society was his belief in alchemy rather than a wish to further the cause of natural philosophy. To this end he threw out a private challenge to the Society's new members that if any one of them could create gold,

13 Sir Thomas Gresham and Gresham College, Bishopgate, London. The college was founded as an alternative place of learning to Cambridge and Oxford, and it was here the Royal Society held its meetings during the seventeenth century. It was also here that Robert Hooke had his apartments and erected in the quadrangle an aerial telescope through which Halley also made observations.

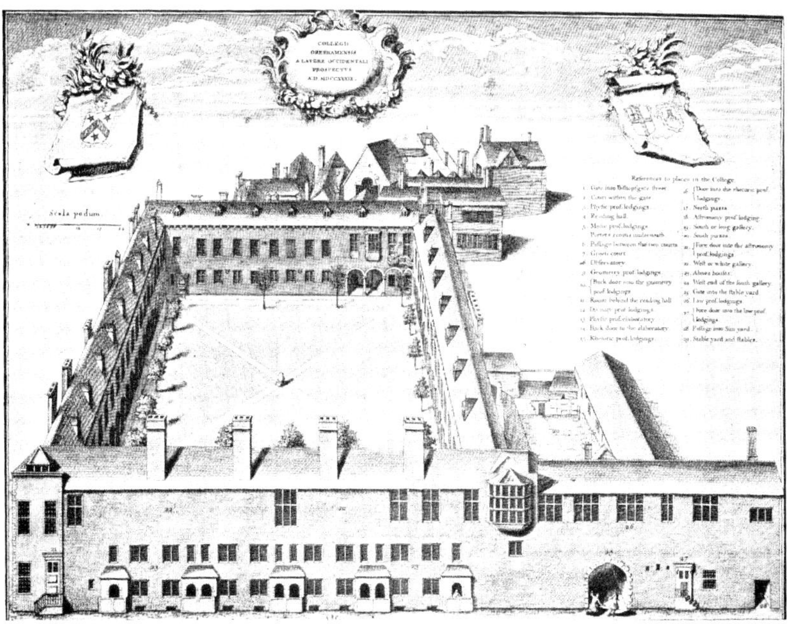

he would promote him to chief minister! Wren, it seems, in soliciting royal patronage for the new Society, did little to dispel the King's ready optimism for new, easy riches to support his court's lavish lifestyle.

From the start the range of subjects discussed at the Royal Society's weekly meeting was very diverse and sometimes trivial in the extreme. With a few exceptions the City's notables and new men of influence were all encouraged to seek membership. One exception was the philosopher Thomas Hobbes, a man of genius but of lower rank with a reputation as a subversive and atheist. Pepys was admitted during the advent of the Plague comet in 1664 while his good friend and fellow-diarist John Evelyn had been there from the beginning. Pepys had no pretence to understanding much science, but he was well read and a collector of books. He thought Robert Hooke's *Micrographia* (1665) "the most ingenious book that I read in my life", but other scientific books were usually hard going for him. John Evelyn, however, was more of a genuine intellectual and had a reputation for precocity. It is said that in his fifth year he was a Latin prodigy and could decline all the nouns, conjugate the irregular verbs and recite an astonishing French as well as a Latin vocabulary. People like Evelyn, with their bias for classical studies, permeated the English governing class. Education and high learning were in the air; it was not an uncommon experience for schoolboys to discuss Caesar's *Commentaries* in Latin, as equals, with their fathers just as father and son today might chat about a favourite TV programme.

CHAPTER 3

Grand Tour and the Comet of 1680-1

After his return from Danzig, Halley soon fell back into his role of Royal Society dilettante. He was still amply supported by his father and was now firmly established among the London virtuosi. However, shortly after his trip to Danzig he grew restless again. By now he was aware that high learning was to be found on the Continent as well as in London. It was now he conceived the idea of an intellectual grand tour of Europe to take in the sights and significant capitals and at the same time note what new science the virtuosi there had to offer. In particular, Halley wished to meet the savants of the Paris Observatory where some remarkable new discoveries were being made.

Then, on 14 November 1680, several weeks before Halley and a companion left for Paris, a new comet was discovered in the morning sky in the constellation of Leo by Godfrey Kirch, observing from Coberg in Germany. This was not the comet destined to carry Halley's name. That was still two years off, but it was a comet whose appearance was later to exercise a powerful influence in his relationship with Newton.

It was the first ever comet discovered telescopically. Once seen, however, it was then faintly visible to the naked eye. Two days later Kirch noted with his telescope the first signs of a rudimentary tail. By 21 November the comet had rapidly increased in brightness and was now equal to the first-magnitude (1^m), orange-red star Regulus near by. By the 24th the tail was plainly visible to the naked eye and on the 28th it had grown to a length of 15°, or about three times the distance between the pointer stars of the Great Bear.

Gradually the comet drew closer to the Sun and then it became difficult to follow in the twilight sky until finally it was lost from view. Shortly after its closest approach to the Sun (perihelion) it emerged again into the early morning twilight, showing a magnificent tail that stretched 70° across the sky.

The comet attracted world-wide attention and became the greatest spectacle anyone had ever seen. In London and Paris, city dwellers who had never before bothered to look up in the night flocked outdoors in their hundred thousands to get a view of it.

Throughout Europe the appearance of the comet created a great furore and the same kind of consternation which had Evelyn in London note in his diary it might well be "a warning from God". There were many who shared similar sentiments and dreaded the new comet presaged terrible calamity. In Germany medallions were struck to counter its effects with the engraved message: 'The star threatens evil; only trust! God will make things good."

One German scholar wrote: "I tremble when I recall the terrible appearance it had on Saturday evening in the clear sky, when it was observed by every-body with inexpressible astonishment. It seems as though the heavens were burning, or as if the very air were on fire ... from this little star stretched out such a wonderfully long tail even an intellectual man was overcome with trembling; one's hair stood on end as this uncommon, terrible, and indescribable tail came into view."

14 A German medallion struck to commemorate the 1680–1 comet.

15 Some notable comets of the past recorded on an old print. (1) Tycho's comet of 1577. (2) The comet of 1680 ("Newton's comet"). (3) Napoleon's supposedly lucky birth comet of 1769, discovered by Charles Messier.

Among the strangest stories concerning the 1680-1 comet is one that tells of a hen in Rome which laid an egg with the image of the comet impressed upon it, and later a German woodcut was provided as evidence of the fact (Fig 16).

Later Pierre Bayle in his *Pensées sur la Comète* relates similar stories. He also tells us the sceptics were having fun at the expense of comets much earlier in the century and cites a remark gleaned from the correspondence of a M de Bassompiere to a M de Luynes shortly after the death of Philip II: "It seems to me," wrote the French wit, "that the comet we laughed about at St Germain is no laughing matter, as it has buried in two months a pope, a grand duke, and a King of Spain!"

Halley, it appears, never held astrological beliefs about comets. One of his new acquaintances at the Royal Society was his biographer John Aubrey—a man thoroughly steeped in the lore of the older astrologers. Aubrey once suggested to the young sceptical Halley he might benefit from a study of the subject via the works of Aubrey's friend John Gadbury.

John Gadbury had contributed his fair share of gloom and doom during the advent of the 1664 and 1665 Plague and Fire comets. Subsequently in 1665 he had published an 80-page work, now very rare, entitled *De Cometis: Or A Discourse of the Natures and Effects of Comets . . . Historically and Astrologically Considered.* Gadbury appears to have had very little idea about comets in general and was so confused he cites the appearance of three comets for 1664/5 (Fig 17). Not surprisingly his book was ignored by Halley and the rest of the genuine astronomers at the Royal Society.

An interesting relic of this sometime friendship with Aubrey is Halley's horoscope drawn up in Aubrey's hand for his *Collection of Genitures.* The horoscope itself was cast either by Gadbury or Henry Coley, the adopted son of William Lilly. (*see* p 13)

Halley was among the first astronomers to resight the comet of 1680 after it passed round the Sun. Its reappearance was unexpected and took all by surprise. At first everyone thought it was a new comet. Bright comets appearing in quick succession were not new events. It had occurred before in 1618 and then in 1664-5. Some supposed it had happened yet again. Halley himself was

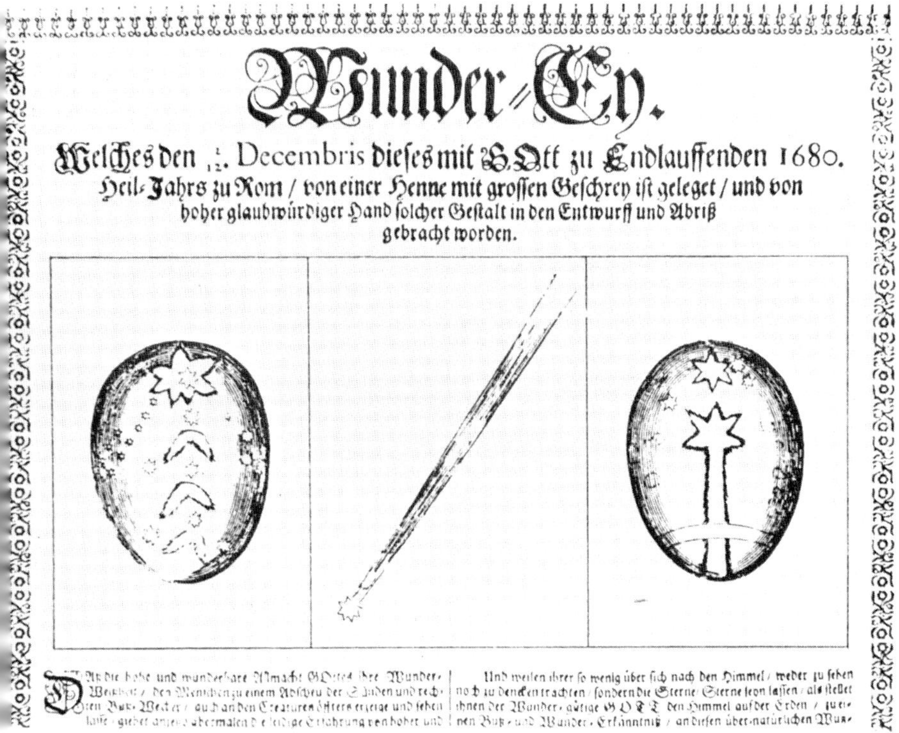

16 An old German woodcut showing the famous impressed egg supposedly laid in Rome after the appearance of the 1680–1 comet. Subsequent to this comet, prints have appeared with Halley's comet identified on similar eggs. However, these must be considered as borrowings by counterfeiters using the original fraud as a basis.

now on the road to Paris at the start of his Grand Tour and was accompanied by his friend Robert Nelson, an old school chum who had been elected to the Royal Society about the same time as Halley. The pair were the same age; both had city-merchant fathers and backgrounds in common.

Their crossing of the English Channel was a particularly long and rough one. Because of the sudden and unexpected appearance of the new comet and its remarkable brilliance, they were anxious to reach Paris as soon as possible to make contact with the savants of the observatory there.

Meanwhile in Trinity College, Cambridge, Newton, who had not yet met Halley, was busy observing the new comet with a monomania for it which matched that he had previously shown for the comet of 1664. Whenever the sky was clear he tracked it and recorded its shifting positon across the background stars.

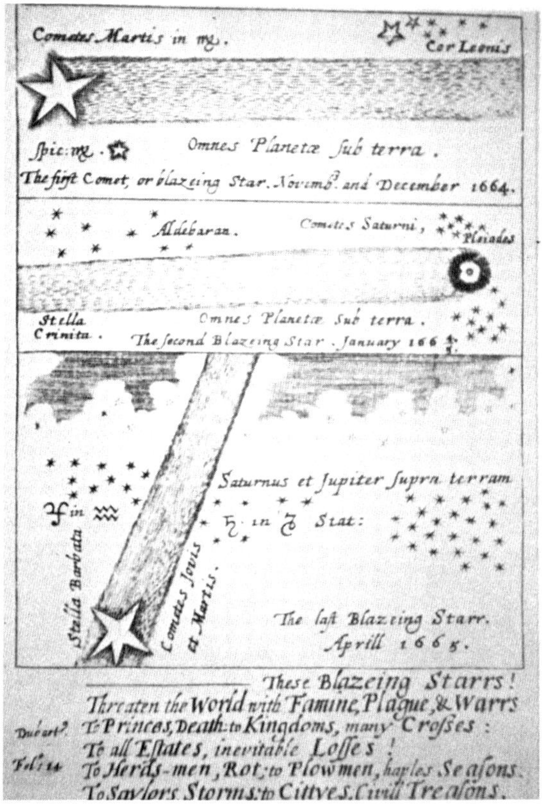

17 The title-page of John Gadbury's book *De Cometis* written about the Plague and Fire comets of 1664 and 1665. The illustrations wrongly cite the appearances of three comets.

Newton is traditionally pictured as a theoretical, rather impractical, chair-bound natural philosopher with little interest outside his study library or laboratory walls, yet his assiduous comet observations reveal him otherwise.

Initially Newton watched what he presumed was a new comet with the naked eye, but because he was shortsighted, he tried using a monocle which made it much more distinct for him. Like everyone else at this time he believed the second comet was a new one—rather than the reappearance of the earlier comet after it had rounded the Sun, for practically nothing was known about their movements. The first had disappeared *towards* the Sun; the new, more brilliant one was apparently moving *away* from the Sun. Soon, however, people like Flamsteed, observing from Greenwich, decided the most logical explanation was that the second comet was in reality only the first retreating after it had approached the Sun and then turned round, but, surprisingly, Newton resisted this idea longer than most and only several years later was persuaded that Flamsteed had been right.

As the second comet began to fade, Newton switched to using a "three-foot" telescope. A week later the comet had grown dimmer and a more powerful telescope was required. He switched to a "seven-foot" equipped with a micrometer eyepiece so he could measure the angular offsets against the background stars. Later he planned to use all these observations to try to calculate the true path of the comet as it curved across the sky. From the very first he had in the back of his mind that this comet might be the very one to shed light on the age-old mystery concerning their movements.

By 10 February 1681 the comet had grown so faint Newton was unable to find it. Nevertheless, he kept up his nightly vigil and finally succeeded in locating it again on 9 March. By then it was extremely dim and had lost all its visible tail. Newton was probably the last man on Earth to glimpse it. Although in modern comet catalogues it is usually listed simply as "the comet of 1680", it is also sometimes known as "Newton's comet".

When Halley and his friend Nelson finally reached Paris, they lost no time in introducing themselves to Giovanni Domenico Cassini, Director of the observatory. This man was to head a dynasty of equally notable astronomers and scientists who only died out with the death of Jean Dominique Cassini in 1845.

Cassini already knew of Halley's growing reputation in British astronomy and offered him open hospitality and the opportunity to observe the brilliant comet using the instruments of the observatory. Halley seems to have been generally disappointed with the observatory's equipment, his expectations in this matter having been too high. Nevertheless, he secured a series of positions for the comet as Newton was doing in Cambridge and Flamsteed at Greenwich. Later, together, these observations provided raw data for Newton's and then Halley's own investigations into comet orbits.

In Paris, Halley also met the astronomers Picard and De La

Hire, the men who two years later were to find another comet which was to shape his destiny. It had been a copy of De La Hire's planisphere, borrowed from Hooke, that Halley had used as a model for his own star map of the southern hemisphere after returning from St Helena.

From Paris, Halley corresponded with Hooke in London. It was Hooke who had now taken the younger man under his wing. Before Halley departed for France, Hooke had requested that he look out some new scientific books for him which were difficult to come by in London; Flamsteed had likewise requested he secure him a new map of the Moon.

In his first letter to Hooke, Halley tells him "... the general talk of the virtuosi here is about the Comet...". In a later letter he relates: "... Monsieur Cassini did me the favour to give me his books of ye Comett Just as I was goeing out of towne ... he has given a theory of its motion wch. is, that this Comet was the same with that appeared to Tycho Anno 1577, that it performes its revolution round the Sun in a great Circle including the Earth . wch he will have to be fixt in about 2 yeares and halfe.,.."

Halley adds it may be difficult for Hooke, who was also much interested in comets, to accept this idea, but he intimates that he himself has some regard for it. Using his own observations, obtained at the Paris Observatory, Halley attempted to reconcile the comet's movement with the idea of a constant rectilinear (straight-line) motion but found this was unsatisfactory—finding its path was neither constant nor straight. Nevertheless, Cassini's own ideas seemed subconsciously implanted in him so that in later years, when he set himself the task of studying comets again, they initially handicapped him.

But comets were not to be the only subject that interested Halley in Paris. Everything about the city and its culture was grist to his mill. Nothing he gleaned went to waste, and many years later he was making use of basic data picked up during his stay in the French capital. His outgoing style and his gift for friendship, as well as his learning, seem to have made a very favourable impression on the savants. Generations later the famous scientist, and sometime comet-watcher, François Arago, who served as director of the Paris Observatory after the third Cassini had

18 The French Academy and the Paris Observatory visited by Louis XIV and his entourage. Halley observed the comet of 1680-1 from here during his Grand Tour. It was also from here that the comet of 1682 (later Halley's comet) was discovered by two French astronomers.

held that post, recalled that the young English astronomer was remembered as "a charming man of rare intelligence".

After Paris, Halley and Nelson toured various parts of France then finally headed towards Rome, stopping over at the other fashionable centres of culture. In Rome itself we find Halley enquiring about Greek and Roman foot-standards and immersing himself among the wealth of antiquities. Like Newton, Halley was much taken by the cultures and learning of the ancient world, and the pair pursued these interests with the same zeal they gave to astronomy.

CHAPTER 4

Changing Fortunes and the Comet of 1682

How long Halley and his companion, Nelson, had intended to remain in Europe after setting out across the Channel in December 1680 is not known, but in November 1681, while still in Rome, Halley received some family news which necessitated he cut short the Grand Tour and return to London.

Very little is known about the circumstances that made Halley abandon his continental plans. It has been suggested it was news of his father's impending second marriage, or perhaps his father insisted his son return to London to attend the ceremony. The date of the second marriage is unknown. However, another significant factor seems to have been the dwindling family resources. Much property had been lost in the Great Fire, and while there was still sufficient income to educate and support his children, Edmond senior's fortunes had gradully declined so that by 1681 they were at a low ebb. It is also possible that Edmond senior, under pressure from his new wife, or his wife-to-be, had told his son he was no longer able to support him in the life of a gentleman.

A clue here is that Halley seems to have disliked his stepmother from the start. Perhaps the prospects of his father reducing his allowance, and the knowledge that in the very near future he would have to find alternate means to support himself, decided him to call a halt to the leisurely jaunt round Europe. Whatever the reasons for cutting short his stay in Italy, his actual return to London was not immediate, and he recrossed Europe at a comparatively unhurried pace, taking in Genoa, then Paris again. He finally reached London via the Hook of Holland sometime in January 1682.

Halley's companion, Nelson, remained in Italy. His stay there had been a fortunate one. In Rome he had met, and then courted, Lady Theophila Lucy, a wealthy widow and daughter of George, Earl of Berkeley. The pair married in November 1682, and we hear little more of them in the events of Halley's life, except it is believed they and Halley remained good friends.

According to Hooke's diary—unfortunately the only reliable source of information about Halley's day-to-day movements at this period—Halley was back in London by 24 January 1682. With talk of love, money, marriage and prospects for the future in the air, it is perhaps no surprise to find our hero soon committed to matrimony himself. His marriage to Mary Tooke (supposedly daughter of the Auditor of the Exchequer), took place on 20 April 1682—a bare three months after his homecoming.

There, unfortunately, the plain fact ends and conjecture begins. Even the profession of the bride's father remains in doubt, for according to records the Auditor of the Exchequer between 1673-98 was Sir Robert Howard (1626-98), Dryden's brother-in-law. While Howard had a daughter called Mary, this is purely coincidence and she is not the same personage. No Tooke (or Tuke, as spellings vary) appears to have held office round this time. But there were Tookes (or Tukes) aplenty in London. One was a famous poulterer, and there is some evidence pointing in this direction; others were printers and publishers. It is possible Mary Tooke was a widow, yet the fact she is referred to in some instances as *Mrs* Mary Tooke is no evidence as spinsters were usually addressed this way (*see* note 3).

How Halley and Mary Tooke met is also unknown. Outwardly there were no reasons for any apparent haste as their first child was not born until several years later. There is a strong possibility the marriage was a runaway affair. A pointer to this is that the ceremony took place at St James's Church, near Aldgate, an establishment famous at the time for several other quick marriages as couples were not required to wait for the reading of banns or a licence. If it were a quick, runaway marriage or elopement, the reason could be the bride's parents may not have approved of their daughter's suitor. Halley was without prospects. We can imagine the situation: the bride's father was probably a man of substance. If he was Auditor of the Exchequer, or holding a similar office, money would be a factor constantly on his mind. It is possible Halley was seen as a penniless gentleman of fortune—someone after his daughter's substantial dowry. And were the bride's parents aware of the coffee-house gossip surrounding their potential son-in-law's adventures on St Helena and in Danzig?

Mary Tooke would certainly bring with her a dowry. In the immediate years following, it is difficult to see from where else the money came for day-to-day expenses in the light of the fact that the Halley family fortunes had dwindled considerably. And five years later, where did Halley find funds to publish and print at his own expense Newton's *Principia*? It did not come from his £50-per-annum salary paid him as Clerk to the Royal Society because we know the Society never paid him out in hard cash.

Mary Tooke was described by a contemporary as "a young lady amiable for the gracefullness of her person and the beauties of her mind". Whatever the reasons for a hasty marriage, it was a love-match which was to span fifty-four years. By Halley's own admission it was one of the saddest days in his life when his wife died in 1736.

Immediately after their marriage the couple settled in Islington, then a rapidly expanding country village just north of the City. Here, free of the smoke and fumes, Halley set up his telescope and sextant and started on a long-term programme of observing the Moon's changing positions during one of its cycles lasting eighteen years. Martin Folkes, who knew Halley at first hand in his later years, wrote: "... neither the domestic cares, nor joys of a happy marriage could diminish his ardour for astronomy and science."

Unfortunately we have no record relating to the young bride's thoughts on this matter, but no doubt she accepted the role of bed-warmer for her often absent star-struck husband. Whatever Halley's passions for the heavens were, they certainly did not prevent his diligent pursuance of more mundane ardours, and the marriage later resulted in three children: two daughters, Margaret and Catherine, who survived their father, and a son, also Edmond, who predeceased his father by a year.

We hear little of the newly married couple in Islington until a few months later. Then in August 1682 another bright comet was sighted in Europe. Flamsteed observed it from Greenwich, Newton from Cambridge and Halley from his domestic observatory. Although this comet did not prove to be as brilliant as the one seen in 1680-1, it nevertheless attracted widespread attention. Evelyn, in his diary, laconically notes for 18 August: 'This night I saw another *Comett* neere *Cancer*, very bright, but streame its tail not so long.' For Halley himself this comet would assume a great importance and provide the linchpin to his own theory of comets—yet this time was still some years ahead.

The bright comet or comets of 1680-1 (astronomers were not yet universally agreed whether it was one or two separate bodies) and now another bright comet in 1682 were sufficient to rekindle a widespread interest in these peculiar wanderers in space. All over Europe telescopes tracked the newcomer's movements from night to night until it finally faded from sight. Probably the last view of it was by an observer at the Paris Observatory on 22 September.

19 Philippe De La Hire, co-discoverer with Picard of the comet of 1682 (Halley's comet). Halley used his planisphere as a model for his own of the southern stars.

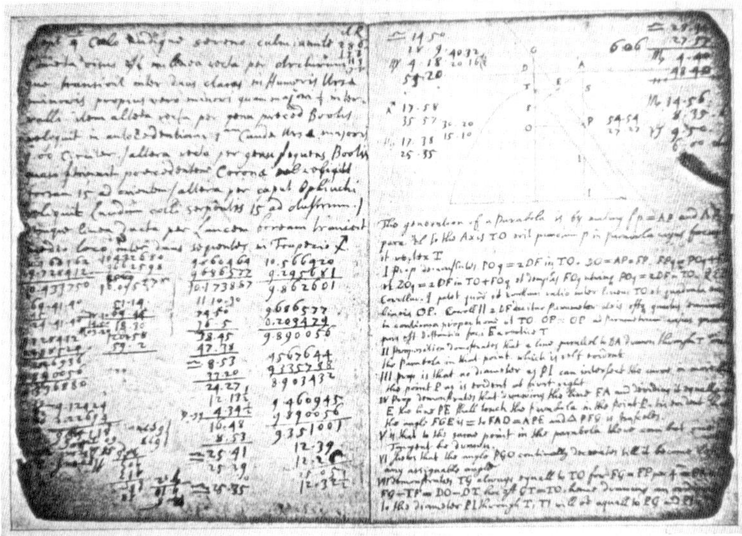

20 Halley's original observations (in Latin) of his sightings of the comet of 1682 (later Halley's comet). The left-hand page is concerned with his reductions for an observation of 4 September 1682. Elsewhere there are observations for seven other nights, but one page is missing.

This volume was probably one of his original college notebooks and is quaintly titled in his hand: "Edmund Halley his Booke and he douth often in it Looke."

The right-hand page concerns itself with the properties of a parabola which *may* indicate that before his contact with Newton, his own thoughts were already focused on the possibility of parabolas for cometary orbits. However, it is likely this juxtaposition is coincidental and the explanation is he simply used the empty pages of his old college notebook to record these (later) observations. The question, nevertheless, remains a very intriguing one.

After marriage, Halley's routine soon became firmly established. The dark hours, when skies were clear, were dedicated to watching the heavens. During the day he would commute to the City and busy himself in Royal Society affairs. All this time Halley had still not met Newton for the Cambridge recluse seldom visited London and the Society.

Halley had no professional occupation and whatever monies, if any, he received by way of an allowance from his father's resources, they certainly must have been supplemented by income derived from his wife's dowry.

Although by 1682 Halley had already made his mark in the field of astronomy, he was only on the first rung of the ladder to

21 Halley's comet as drawn by Hevelius in September 1682. The head shows a remarkable flame-like structure reminiscent of the form later seen during the comet's 1835–6 apparition.

fame. The way all scientists made a claim to fame was to make discoveries and then write papers for publication about them or, alternatively, to write learned discourses about others' discoveries and shed further light on the subject. In the year following the 1682 comet, Halley published two scientific papers: one about the motion of a satellite of Saturn, which had been discovered by the Dutchman Huygens in 1655, and the other entitled "A Theory of the Variation of the magnetical compass". During his lifetime, and for some little while after, Halley's contributions to the science of the Earth's magnetism, nowadays called geophysics, was considered more important than his researches into comet orbits and his other astronomical achievements.

It was in the spring of 1684 that Halley turned his attention to the famous three laws of planetary motion which had been formulated earlier in the century by Johannes Kepler, Tycho Brahe's protégé. The discovery of these laws by Kepler was considered to be a break-through in astronomical knowledge almost equal to that of Copernicus when he set forth his ideas about the heliocentric nature of the solar system. However, Copernicus and his contemporaries were unable to measure the position of the planets with any great accuracy and consequently assumed that the planets moved round the Sun in perfect circular orbits. Whereas observing errors of a half a degree in position might have occurred in Copernicus's day, by the end of the sixteenth century Tycho Brahe had reduced errors down to one or two minutes of arc.

Kepler inherited Tycho Brahe's painstaking observations and decided to use them to draw up tables for planetary motion. His investigations into the orbit of Mars had started while Tycho was still alive, and he had boasted he might solve the matter in eight days... It took him several years. When he started, he was under the full and firm conviction that the age-old idea of beautiful and simple circles, an idea which held sway over the human mind for more than 2,000 years, was unquestionably true. His task, then, was simply to frame a combination that would account for some new, inexplicable anomalies in the motion of Mars—if Tycho's observations of the planet could be trusted.

The amount of industry, sagacity, perseverance and inventive genius displayed by Kepler is almost unparalleled in the history of astronomical discovery. To exhaust the subject and to prove that no combination of circles could possible represent the exact movements of Mars, he enumerated nineteen different hypotheses spread over 900 clearly written folio sheets of calculations. He realized there could be no question that Tycho's observations contained errors to mislead him. Having adopted a hypothesis, he laboriously computed what ought to be the visible position of Mars, as seen from Earth, throughout its entire revolution. Then he compared these *computed* places, or positions, with *observed* places—those actually occupied by the planet.

By the standards of the day the discrepancies he found

22 (*a*) Johannes Kepler (1571–1631). (*b*) A diagram reproduced from Kepler's book *New Astronomy* (1609) demonstrating the movements of Mars

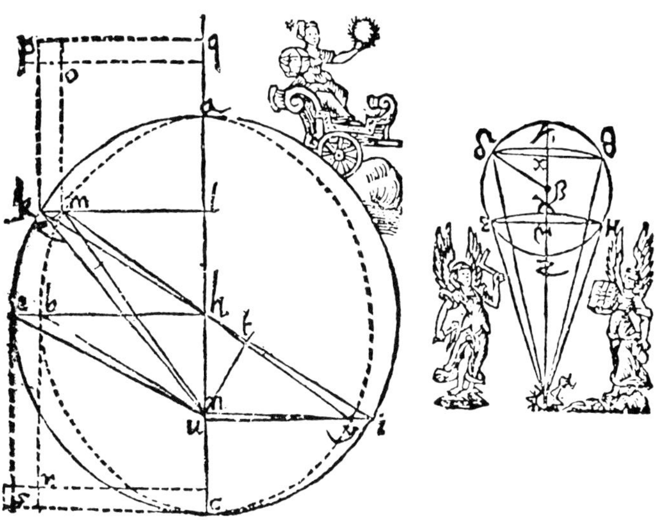

between *observation* and *computation* were not large. They did not exceed a tenth part of the Moon's apparent diameter, or about 3 mintues of arc. Yet on this slender basis Kepler still kept faith with Tycho's observations of the planet. Others would long ago have thrown in the sponge and taken to working on other less intransigent problems. Not so Kepler: if needs be he would reconstruct an entirely new science for the heavens to explain the puzzling anomaly.

Nevertheless, the demonstration of the great negative truth, i.e., that the perfect circle, with all its beauty, classical simplicity and its strong historical antecedents dating back to Greek times, was an erroneous concept was a bitter pill to swallow. Yet its final acceptance by Kepler, as far as the planet Mars was concerned, was the first step in creating the new order. Common sense reasoned: if not a circle, there must be another curve which might prove to fit the motion of Mars. Indeed there was: an oval—or rather the ellipse which instead of one central point has *two foci* from where lines drawn to some point on the curve are, when added together, equal in length to the major axis (Fig 23). This idea provided Kepler with what was to become his first law: *Every planet revolves in an elliptical orbit about the Sun, which occupies one focus.*

During the course of his researches, when the possibility of oval-shaped paths for the planets and the Moon was first considered, it had occurred to him that if this were correct, it would follow, as a consequence, that a planet's or a satellite's speed would not be constant and would depend on what part of its orbit it was moving in. This insight proved correct, and he now realized it provided him with a second law: *Any planet or satellite during the course of its path sweeps over equal areas in equal times* (Fig 24).

Kepler's discoveries were considered to be a triumph over nature. Nature had been compelled by man's persistence to render up two more long-inscrutable secrets. The prejudices of two millenniums had been swept aside just as Copernicus had done three-quarters of a century earlier. Kepler was now sure that all the other planets and satellites in the solar system obeyed the same dictates. Even so his work remained unfinished. With the acceptance of the elliptical theory, it became apparent to him there existed too many common points of similarity between the revolving planets for them not also to involve some common bond which linked them into an all-embracing scheme of mutual dependence. Kepler was sure another of nature's inscrutable secrets still lay hidden.

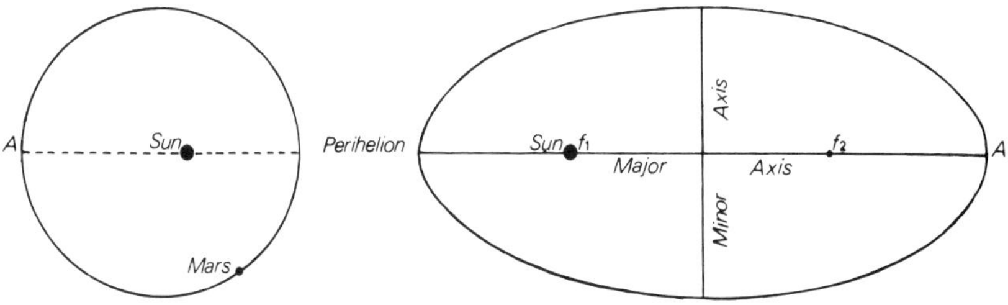

23 The orbit of Mars and the properties of an ellipse. (*right*) The eccentricity is measured by expressing the distance between the two foci f_1 (Sun) and f_2 in terms of the length of the major axis. (*left*) Kepler discovered that the planet Mars moves in an ellipse, but compared with the very elliptical orbits of comets the orbit of Mars has very little eccentricity, and on this scale the two foci are positioned very close together.

To outsiders, Kepler's new task to find the unknown linking factor looked hopeless. This new problem occupied and exercised his genius for a further ten years. During that time he eliminated in turn those factors he believed not significant in solving the puzzle. He finally selected the *mean distances* of the planets and their *periodic times* as the two elements for special

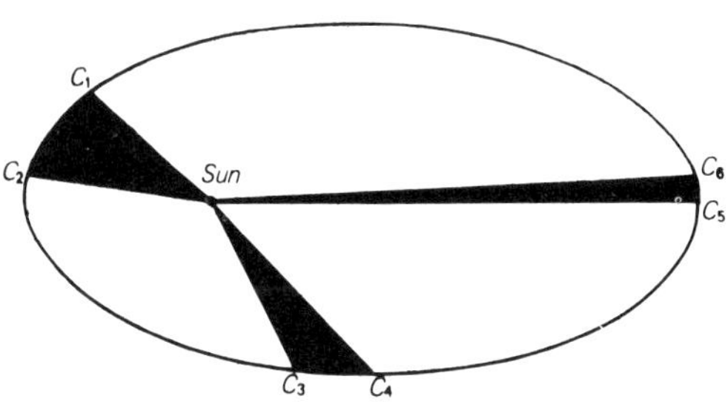

24 Kepler's second law (of equal areas). A comet (or any body in elliptical orbit) takes the same time in moving from C_1-C_2 and C_3-C_4 and C_5-C_6. The swept area $C_1\text{-}S\text{-}C_2 = C_3\text{-}S\text{-}C_4 = C_5\text{-}S\text{-}C_6$.

investigation. He noted, particularly, that if a planet were more remote from the Sun, its period increased but not in the exact ratio of the distance. Finding that no simple proportion existed, Kepler turned away from ratios involving geometry which till then had almost exclusively been employed in astronomical investigations. It was now he conceived the idea that the undiscovered secret might be found in proportions existing between some arithmetical powers of quantities. First he tried the *squares* of the periods and distances; again he was unsuccessful. Then in turn he tried the *cubes*, but as in the case of the squares no proportion was discovered to exist among the third powers.

He almost gave up in total despair, but then, after further reflection, he had an idea. What if the inscrutable was a combination of squares *and* cubes? Nature had already shown that the simple circle was inadequate to explain the complicated paths of the planets round the Sun. Was she even more perverse in arranging the number relationships that existed between the planets? Indeed she was. It was another cry of Eureka! two thousand years on: *square* the periods of revolution and *cube* the mean distances and the great secret at last lay revealed.

The third law now joined the first two: *The squares of the periodic times are exactly proportional to the cubes of the mean distances*. Kepler soon confirmed that the satellites (a name he coined himself) of the planets also revolved round their principal in obedience to its sway. With these new laws of nature firmly established, dynamical astronomy entered a new age (*see* note 4).

Kepler just missed discovering universal gravitation for himself. He suspected his three laws were but corollaries of an even higher law remaining to be developed. But he could go no further, for even Kepler's mind was steeped in medievalism and programmed to the traditions of the past. He also missed discovering that the comets as well as the planets might obey his three great laws. Apart from one man in England who hit upon the truth of it, but whose suggestion lay forgotten, it needed the enlightened men of later generations to reconcile comets with Kepler's laws—men who were able to look beyond the old limited horizons.

When Halley pondered over Kepler's laws in 1684, he was

puzzled about certain consequences following on from them. He wondered, for example, why it was that the time a planet takes to travel round the sun should always relate to the ratio 2:3 as in the third law.

Soon he realized that this ratio might be a consequence of the Sun's attraction on each planet—an influence, or force, which diminished by *four* times if the distance were doubled, or *nine* times if the distance were trebled.

Even Halley's blossoming talent was not enough to help him discover a mathematical proof of it. Neither was Halley the first to think about the Sun's attractive force in relation to Kepler's laws. He realized this when he raised the matter with Hooke and Wren at the Royal Society. He found the notion of such an attractive force had been in the air with several of the virtuosi for more than a decade, but no one had been able to frame it in concrete terms, i.e., mathematically. Nevertheless, to both Halley's and Wren's surprise, Hooke now claimed to have worked out a geometrical proof of it. Pressed to show it by the other two, he declined their invitation. Wren more than Halley was immediately suspicious of Hooke's glib off-the-cuff disclosure. He and Hooke had been associates now for many years; he knew Hooke's sometime boastful character better than anyone. For this reason Wren decided to put him to the test by suggesting an open competition. If either Hooke or Halley could prove the proposition, he would present the first to submit it for scrutiny with a book of his own choosing to the value of forty shillings.

The prize itself was a trifle, but at the same time it was a direct challenge to Hooke. Wren had called his bluff. Halley himself now fascinated by the problem tried again to find a proof but soon admitted to Wren he was still defeated by it. It was now up to Hooke to come forward and claim the prize. Caught out by Wren's ploy, he dithered. He told the other two he had now decided to keep his own proof a secret until others had had a further chance to work it out. By doing this he claimed they would value his solution even more when he presented it.

Hooke was almost certainly playing for time, hoping meanwhile to hit upon a solution and save face. Four years earlier he had written to Newton about his attraction theory remarking:

"...my supposition is that the Attraction always is in a duplicate proportion to the Distance from the Centre Reciprocall and Consequently that the Velocity will be in a subduplicate proportion to the Attraction and Consequently as Kepler supposes Reciprocall to the Distance."

Newton did not reply!

Hooke could not guess what young Halley's next move might be and its momentous consequences. The young astronomer was no face-saver himself: he was a seeker after truth. He was not too proud to admit defeat and seek assistance. Perhaps only one man in England could help him. A little earlier they had opened a correspondence for the first time when the Lucasian Professor of Mathematics at Trinity College, Cambridge, had written to him in London requesting his observations of the comets of 1680-1 and 1682. Yes, perhaps Isaac Newton might be the very man to help him develop his own proof...

CHAPTER 5

Prelude to Principia

Gravity was no new force discovered by Newton. The idea of a force acting on the Earth and on the other planetary bodies had been around long before Hooke and others inferred that Kepler's laws may provide insights into it.

Back in Greek times, Aristotle had laid down the law of falling bodies attracted by the Earth's pull. It asserted that in the case of balls of unequal weight which were dropped at the same instance, from equal height, the velocity of the two would be directly proportional to their respective weights. The idea sounds logical; it sounded so logical that no one put it to the test until Galileo experimented 2,000 years later, perhaps, for convenience, using the Leaning Tower of Pisa as legend has it.

Galileo proved Aristotle's idea false. His experiments showed that the velocity of a ball (or any other object) or the time taken in the fall was, neglecting any air resistance due to our atmosphere, entirely independent of the weight of the ball. And who can forget the repeat of Galileo's famous experiment demonstrated for TV audiences three centuries later on the airless surface of the Moon during the historic Apollo 11 mission in July 1969.

Galileo developed more experiments and theories to show how objects moved in dictate to the Earth's gravity field. He proposed that bodies set in motion will move off with uniform velocity in a straight line—sustaining the direction in which the impulse is applied. Such bodies he maintained would continue to move for ever unless some force or influence is exerted to change the direction or destroy the velocity.

This could not be proved directly from experience, but its concept was generally accepted by others as the first law of motion. Previous to this, human imagination seems not to have

25 While Galileo Galilei (1564–1642) did not further knowledge of comets directly, his experiments nevertheless foreshadowed some of Newton's laws of motion which led to the understanding of how comets move.

conceived the idea that motion could continue after the cause producing it had ceased to act. Yet, ironically, there is no motion produced by human contrivance as, for example, the flight of a stone from a sling or a cannon-ball from a gun where it is not apparent that the force producing the motion stops its action after the first impulse. Neither the sling nor the explosive charge with its expansive power cannot pursue the stone or cannon-ball once liberated.

Galileo went on to discover a second law of motion whereby rectilinear (straight-line) motion changes into curvilinear motion. For example, if a cannon-ball is projected horizontally, it soon begins to fall away from the straight line in which its flight began. Galileo proved that under the united effect of the projectile force, and the force (gravity) which caused it to fall towards the Earth, the ball would trace out a regular curve called a *parabola*, a shape similar to an ellipse but whose major axis is infinitely long (Fig 34).

In his experiments to show the falsehood of the old Aristotelian doctrine, Galileo was careful to point out that while two unequal bodies fell with equal velocities and struck the ground at the same instant, the effect of the blow struck by the heavier body was very different from that produced by the lighter one. He demonstrated that the effect was precisely proportional to the weight of the falling body and that, for example, a body ten kilos in weight would strike a blow ten times more powerful than a one-kilo weight after falling through equal heights. Thus in effect a blow struck by a moving body must take into account not only velocity but also the weight, or mass, of the body. Mass multiplied by velocity is the *momentum* and a true representation of the effect produced by a blow struck by a moving body.

In formulating his new concepts about mass and motion, Galileo foreshadowed Newton's first two laws, but much like his contemporary Kepler, Galileo was still shackled in medieval ideas and was not able to make the necessary mind-step made by Newton, born the very year Galileo died. Kepler on the other hand, had a little more imagination than Galileo. He had felt the necessity to discover and reveal the constantly active power which operated in the direction of planetary motion—believing as he did that without some ever-active force, the planets must of necessity stop in their tracks. In contrast to Copernicus who sought merely to *describe* order, Kepler sought the *reason* for it. Yet it was Kepler's ignorance of the laws of motion that caused him to miss the discovery of universal gravitation and leave it as an unripened fruit to be plucked from the tree of knowledge by Newton.

Following Kepler's lead, others in the first half of the seventeenth century attempted to flesh out his ideas to obtain a better understanding why the planets moved as they did and why they obeyed his laws, but they were men of a lesser genius and no one achieved anything satisfactory. Nevertheless, it was clear that all heavy bodies were in some way attracted by the Earth—this was apparent from the everyday fact that every body falling freely sought the Earth's centre. But the very nature of this central force lodged inside the Earth and other bodies, and how it could operate unseen, remained one of God's apparently inexplicable mysteries.

The ancients had noted that when a stone in a sling is whirled rapidly round the head of the slinger, a force was developed which stretched the string. This force was later defined as the *centrifugal* force (a word coined by Christiaan Huygens, the Dutch physicist) and it came to be accepted that this force was probably significant in keeping the planets in their orbits round the Sun. Thus a tendency to fly from the centre must somehow be generated in the planets revolving in the solar system.

To account for celestial motion the theory that was widely accepted was put forward by the French philosopher René Descartes (Latinized Renatus Cartesius). This was the all-embracing Cartesian theory published in 1644 in which he imagined space to be filled with a thin fluid matter, but here and there were centres where this fluid matter formed vortices. In the solar system there were supposedly fourteen such vortices. The largest contained the Sun, and then there were vortices containing the chief planets, the Moon and the four (known) moons of Jupiter and (later) the two (known) moons of Saturn. The fixed stars, when they became opaque, ceased to be centres of vortices and passed from the vortex they formerly inhabited to the vortex next to it. From this vortex they passed to the next ad infinitum. It was these wandering stars which Descartes believed were comets.

Each planet with its own vortex was embedded in a larger vortex that whirled about the Sun. The presence of these various whirlpools was the very reason for gravity. Thus, according to Descartes, bodies fell to the Earth because they were caught in the Earth's vortex Q E D. The Cartesian theory became the dogma of the middle and late seventeenth century. For several generations following Newton's *Principia* it still had its supporters even in England.

During the middle seventeenth century one man firmly on the right track to finding a solution to the problem of planetary stability was the Italian Giovanni Alfonso Borelli, who combined his scholarship in mathematics and astronomy with medicine. About 1666 he accounted for the pulls of the planets in terms of two forces: a force urging a planet to unite with the Sun and a (centrifugal) force urging the planet away from the Sun.

26 The ideas of René Descartes (1596–1659), latinized Renatus Cartesius, dominated scientific thinking in the period immediately before Newton. His vortices theory (whirlwinds as Voltaire later disparagingly called them) was generally believed to explain the movements of the planets and comets.

By the interplay of these forces, Borelli tried to account for the fact that a planet's orbit (as demonstrated by Kepler) is not a circle. Unfortunately, like Hooke, his admirer, he was not able to match his descriptive insight with the necessary mathematics to prove it.

Huygens solved the problem of the (centrifugal) force developed by a body revolving in uniform velocity in a circle, but this solution could not be applied to the elliptical motions of planets. Both Borelli's and Huygen's work was well known to Hooke and Newton, and Newton had Borelli's book in his library. Hooke tried to extend Huygens's work by experiment, but this led to no definite result. In 1674, however, he showed that he had come as near to a solution as was possible without using mathematics. He believed there were three assumptions to be made. Firstly, every heavenly body attracts to its centre every other heavenly body. Secondly, every moving body moves in a straight line until deflected therefrom by another force. Thirdly, the attraction between two bodies is greater the closer they are together.

Hooke claimed (rightly) that all heavenly motions could be explained by these assumptions; he also claimed he did not have the time to pursue the matter and perform the calculations to prove his ideas. The fact was he had gone to the limit of his intellectual ability; he could not foresee that Newton, whose Theory of Colours he had criticized in 1672, would soon take up the challenge.

While the idea of some kind of dynamical equilibrium between the Sun and its attendant planets, and between the planets and their satellites, seemed to be firmly established by the early 1680s, the nature of gravity itself was a great puzzle. Kepler had reasoned the Earth's and the Sun's gravity was in some sense "magnetic". He supposed that the force needed to keep a planet moving must act tangentially along its orbit. He grasped the concept that the centre about which a body revolved must itself be a body and not an empty point (focus) in space. It is perhaps unfortunate that Kepler had studied William Gilbert's epoch-making book *De Magnete* (1600), for this seems to have coloured his ideas. "Gravity," Kepler wrote, "consists in mutual corporeal affection between cognate bodies towards their union or connection (the magnetic property is also of this order), so that the Earth attracts a stone much more than a stone seeks the Earth."

Descartes in his vortices hypothesis certainly grasped the idea of inertia (later incorporated by Newton in his first law), i.e., each particle of the vortex would tend to travel away along the tangent to its circuit round the Earth and in doing so would recede into space. Nevertheless, Cartesian theory led down a blind alley in the understanding of gravity itself.

Another theory was that forwarded by Isaac Vossius, a canon at Windsor who believed gravity might be explained by the diurnal rotation of the Earth. Halley himself rejected this theory, realizing that any rotation would have the opposite effect.

Flamsteed was another who invoked magnetism as the cause of gravity. He supposed comets in particular aptly demonstrated gravity's attractive and repulsive forces. For example, the 1680-1 comet he believed approached the Sun owing to the attraction force but on nearing it encountered the repulsive force which then drove it back. Thus for the wrong reason he correctly supposed the November comet of 1680 was the same as that of December 1680 in a different mode. Ironically, it was Newton's refutation of Flamsteed's magnetic theory for the 1680 comet which caused him to hesitate in accepting that the two apparitions were due to a single body.

When, in 1684, Halley after consulting Hooke and Wren, finally decided to seek advice from Newton, he had no inkling his journey to Cambridge would result in influencing the whole history of science thereafter. It was to change the course of Newton's life as well as his own.

The journey from London to Cambridge circa 1684 was an exhausting full day's ride by coach or on horseback and not undertaken lightly. Halley mentions his August visit in a letter written some years later, but the best account comes second-hand from Newton's own recollections through a close friend, the French mathematician Abraham De Moivre: "In 1684 Dr Halley came to visit him [Newton] at Cambridge, after they had been some time together, the Dr asked him what he thought the Curve would be that would be described by the Planets supposing the force towards the Sun to be reciprocal to the square of their distance from it. Sr Isaac replied immediately that it would be an Ellipsis, the Doctor struck with joy & amazement asked him how he knew it, why saith he I have calculated, it, whereupon Dr Halley asked him for his calculation without any father delay, Sr Isaac looked among his papers but could not find it, but he promised him to renew it, & then send it him."

Halley had no reason to suspect Newton was bluffing as Hooke certainly had been. He already had a formidable respect for Newton as a mathematician and all his life was much in awe of the withdrawn genius who was fourteen years his senior. In personality the two were in sharp contrast. Newton by now had gained a reputation as a dour, introverted man who seldom laughed or even smiled; a man with absolutely no sense of humour. Halley, by comparison, was an extrovert of cheerful

27 Isaac Newton painted by Sir Godfrey Kneller shortly after the publication of the *Principia*.

countenance and always open and direct in conversation and dealings with others; he thrived in the atmosphere of the London coffee-house and was an excellent raconteur. Newton, long suspicious of all his fellow creatures—*especially* his peers at the Royal Society—was apparently instantly disarmed by the young astronomer's candid, friendly style. Nevertheless, the claim that he could not immediately put his hand to his calculations was probably a subterfuge in order to gain time to think. Halley, whatever his face value, was practically a protégé of Hooke, his arch-enemy. Newton would ponder: was Halley in Cambridge on Hooke's behalf? For good reason, Newton was now wary of any of Hooke's small but influential circle of friends. Yet, Newton would also recall that when he had corresponded earlier with Halley over the observations of the comets of 1680-1 and

28 The earliest known portrait of Halley, in his late twenties or early thirties. It was painted by Thomas Murray, but the inscription is not contemporaneous.

1682, there had been the same immediate friendly but respectful tone in the young astronomer's letters. And why, if he were an agent for his enemy, would he reveal to him Wren's suspicions that Hooke himself was playing for time. Perhaps through Halley the good Lord had delivered the old enemy into his hands! He had never forgotten the slight by Hooke when he had criticized his Theory of Colours.

Hooke, true, had since tried to make it up to him, but in replying to his letters, Newton carelessly made a wrong assumption over a technical point Hooke had raised. Hooke, with apparent glee, pounced on this faux pas and had again taken opportunity to criticize him publicly at the next meeting of the Royal Society. This was contrary to the tacit agreement between them that any differences of opinion would in future be conducted in private.

Newton was furious that Hooke so readily dishonoured their compact, but he was equally furious with himself for committing a careless error. If what Halley had told him was true and Hooke was unable to prove an attractive force varies inversely as the square of the distance, in an ellipse, the tables could be turned on him to good effect.

How much Newton's pent-up hatred of Hook actually influenced his actions after Halley departed for London must remain conjectural. During his absence from Cambridge at the time of the Plague, almost twenty years before, Newton had pondered deeply about gravity, but his interest in the subject languished when he moved on to ideas about theories of light and alchemy. Ironically, it was the very correspondence instigated by Hooke in 1679, when Hooke caught him out, which apparently rekindled Newton's curiosity about gravity and planetary orbits.

According to De Moivre's account of events, after Halley departed, Newton immediately repeated the demonstration he had just claimed to have accomplished to Halley, but this time it did not work out! He was now grateful for his over-cautiousness with the young astronomer. He had almost made a fool of himself again and played into Hooke's hands. Yet, as with his earlier faux pas, it was all due to simple carelessness. He soon located the source of the trouble: a hastily drawn diagram had led him to confuse the axes of the ellipse with conjugate diameters. By indulging in his lifelong habit of burning the midnight oil when any problem beset him, he eventually arrived at a correct solution.

The manuscript reached Halley two months later by the hand of Edward Paget (*see* note 5). To Halley's amazement it was not just the simple proof of the problem he had discussed with Newton but a complete treatise consisting of nine pages titled in Latin: *De motu corporum in gyrum* (*On the Motion of Bodies in an Orbit*). As a mathematician of considerable ability himself, Halley at once recognized the sheer brilliance of Newton's treatment. It confirmed what he and others had long suspected: the author was a man whose mathematical powers knew few equals.

Not only did the treatise demonstrate the problem involving Kepler's third law which Halley had raised with Newton but the second and first laws as well. While it still fell short of the complete solution to planetary motions round the Sun, it was the greatest leap forward since Kepler's own work on the subject earlier in the century.

It was now Halley made a second trip to Cambridge to offer Newton his congratulations. This was in November 1684 when Halley persuaded Newton to expand his ideas further and write a complete book on the subject. Newton, flattered by Halley's attentions and unstinting praise, readily agreed to do what had been in his mind already since after the August visit. He told Halley he would compose a tract entitled *De motu corporum*. The task was to occupy Newton to the exclusion of all else for eighteen months.

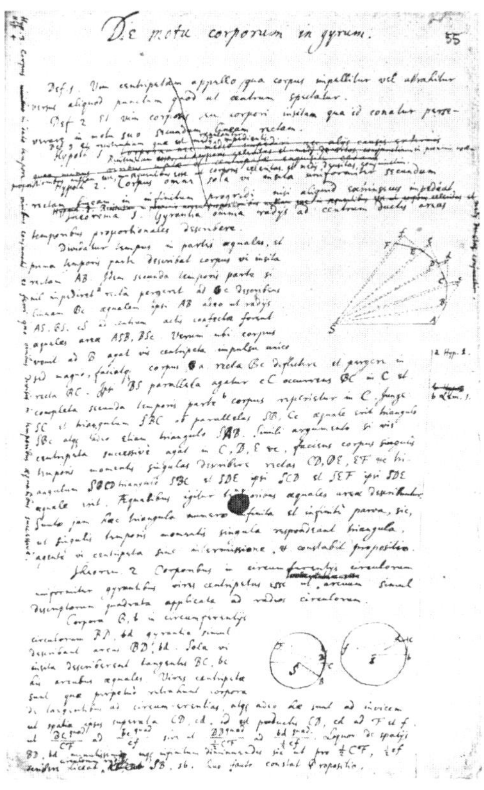

29 The first folio of Newton's tract *De motu*.

CHAPTER 6

Clerkship and Celestial Mechanics

During the writing of the *Principia*, Halley's career was to take a change of direction. Since his father's remarriage, the family fortunes had dwindled even further. Then in March 1684 Edmond senior died in mysterious circumstances. The story relates he left his house telling his wife he would be back by evening, but he was never seen alive again. His body was not found until several weeks later. All knowledge of this incident is derived from a broadsheet published about 17 April 1684, entitled "A true Discovery of Mr Edmond Halley of London Merchant Who was found Barborously Murthered at Temple-Farm, near Rochester in Kent."

Whether it was murder or suicide is conjectual. In view of the fading fortunes, suicide is a strong possibility. Bizarre gossip in later years was to link the astronomer's name with the death of his father, and Thomas Hearne went as far as to declare in his diary: "Dr Halley's Father went in fear of his life from his own Son."

The story is undoubtedly malicious; nevertheless, by this time it is likely that the astronomer's relations with his father were very strained. It seems that the remarriage had caused a breach between them, and there is little doubt at the root of it was the new wife (and money, or rather the lack of it), for it is clear Halley had little time for his stepmother. She married again and reappeared as either Mrs Chester or Cleator—the name is unclear in records. Halley later took his ex-stepmother to court owing to Edmond senior dying intestate. On the evidence of this lawsuit it seems probable that Halley must have continued to derive some small income from his father after returning from Italy in 1682.

Whatever the exact circumstances of Halley's patrimony, and these are not known with any certainty, he was a very eager applicant for the post of Clerkship to the Royal Society when this new post was created in 1685 at a stipend of £50 per annum.

The new post arose as a consequence of the sudden and unexpected resignation of one of the Secretaries which left the affairs of the Society in confusion. The Council, determined that such a situation should not arise again, decided to appoint a paid clerk ("or amanuensis", to quote Halley's own words)—someone who would be answerable to their authority and who would take over the workload of correspondence from the Honorary Secretaries.

It is perhaps strongly indicative of Halley's pecuniary circumstances that he put himself forward as a candidate for this potentially dogsbody post. It required giving up his fellowship, since paid employees were not permitted this distinction, and the Royal Society charter actually forbad it.

There were other applicants for the post, and the voting went to a second ballot. This was held on 27 January 1686 when Halley secured a majority and was afterwards sworn into office. Conditions of employment were that the successful candidate should be a bachelor and live in at the Royal Society's meeting-place at Gresham College. In Halley's case this stipulation was waived, and what influenced the Council's decision was that the young couple had already moved house into the City from Islington to Golden Lion Court in Aldersgate the previous year.

The stipend of £50, even by the standards of the day, was a very modest one. Nevertheless, in the same period some curates were in receipt of incomes only half this, and there were professorships worth only a £100 or so. If not exactly a bread-line salary, it was by no means generous. Pepys at this time had an income amounting to several thousand pounds a year and so did Evelyn. In spite of the poor stipend, Halley may have derived satisfaction he was now earning some kind of living for himself in an environment that suited his talents. With gusto he immediately immersed himself in the Society's business at Gresham, and, reading between the lines, we can detect a resolution on his part that the Clerkship was not to be that of an inferior dogsbody at the beck and call of Council and fellows but rather that of a professional administrator.

First task was to clear up arrears in correspondence; then he took a hand in the meeting-day agenda. Halley's interests were polymathic like those of many of his contemporaries. The tradition of the Society was to discuss diverse and wide-ranging topics—topics quite outside those discussed by scientific societies in modern times.

Even in the seventeenth century it seems Parkinson's law was operative. Employed to relieve the burden from the shoulders of the Honorary Secretaries, Halley generated so much additional work he was hard pressed to fit it all in. He became so bogged down in answering the Society's correspondence he had to apologize to close friends for being dilatory in replying to their personal letters "for want of an Amanuensis to ease me the trouble of copying [the original minutes of the Society's last meeting]".

For someone who himself had recently been appointed to act as an amanuensis, this was a development guaranteed not to endear him to some members of Council.

But other troubles were brewing for the Society's Clerk. Halley's visits to Cambridge had fully awakened the sleeping giant from his intellectual slumbers. For twenty years Newton had brooded about his great works—dating from the time the Plague had closed down Cambridge University and he had returned to his native Woolsthorpe for philosophic meditation.

Newton's notions of universal gravitation did not stem chapter and verse from the time he had watched the famous apple fall in his own orchard sometime in 1666. What seems certain is the falling apple channelled his thinking; but not even Newton's particular brand of genius was able to unravel universal gravitation in extensio without a struggle—a struggle lasting on and off upwards of twenty years. The problems involved were baffling even to his unique mind and they were wrought with overwhelming complexities. Like Kepler before him, he recognized purely descriptive laws were inadequate. Like Kepler he set himself the task of discovering the archetypal laws which in the mind of God presided at the creation of the world; and for Kepler and Newton these laws perforce had to be geometrical/mathematical ones.

Newton's laws took several years to think out and then detail into shape; they involved thousands of hours of thought drudgery. Some of the ideas that finally found light of day in the *Principia* had remained at the border of Newton's consciousness for years, not yet properly grasped or formulated. During the interim his wide-ranging mind was exploring other problems in mathematics, alchemy, optics and in biblical chronology and theology. When, after Halley's second visit to Cambridge in November 1684, Newton had agreed to write his tract, it was accepted as a self-challenge—a commitment to focus all those half-digested ideas and finalize them in print. Halley was the catalyst in the transmutation from idea to reality; he had raised a question at a time when Newton was receptive to it and eager to achieve order from chaos in his vast, yet uncatalogued, storehouse of thoughts. Nevertheless, the knowledge that here was an opportunity to best Hooke must have been added incentive to what he knew would tax his intellectual strength to its uttermost. This view is confirmed from Newton's own lips. When in later life pressed to relate his true motivation for creating the *Principia*, he told his hearer it was really to correct Cartesian errors, but he added that Hooke's past animosity toward him also served to goad him on.

Hooke had been working on the problem of orbital forces since about 1666. Borelli's ideas about the probability of two forces at work governing the motions of the planets had especially influenced him. Although Newton was also later to acknowledge Borelli as a predecessor, it seems unlikely that the grand Newtonian synthesis would have been delayed had Borelli never existed. As a disciple of Galileo, Borelli had wished to develop a rational for celestial mechanics. He rejected the ideas of Kepler's magnetic force. For Borelli, gravity was a constant force and the key to understanding sidereal motions. Like Kepler, he asked the question why planets moved and why they kept their places. Unlike Galileo he accepted Kepler's ideas of the ellipticity of planetary orbits. Both Galileo and Descartes had ignored Kepler's epic discoveries; Borelli tried to correct this. He approached to the very brink of a great discovery, but he was intellectually ill-equipped to make it.

Hooke, taking his cue from the Italian, realized that a planet moving in its orbit is under a constant force pulling it sunward. This, he concluded, was the reason it does not move off in a straight line and disappear into outer space. By 1670, Hooke had also come to realize the force exerted by the Sun was linked with terrestrial gravity. Even earlier he had tried to establish by experiment that the Earth's force of gravity decreased with increasing height above the Earth's surface. It led him to adopt the hypothesis, based on Kepler's third law, that gravitational attraction varied inversely as the square of the distance from the attracting body. Gravity, therefore, Hooke realized, was a cosmic (universal) force and not limited to the Earth. But proving, *mathematically*, how it affected the various elements in the cosmos was quite another matter. This way Hooke quite independent of Newton outlined universal gravity *but only as an idea*: gravity as both Hooke and Newton understood it was a cosmic agency operating between all bodies of the solar system and was the universal force which kept the planets revolving in closed orbits round the Sun.

Hooke's failure to capitalize on his ideas was due to his lack of mathematical expertise; because of his excessive pride, and jealousy of others' abilities, he would not admit this to his circle of peers at the Royal Society.

Had Newton not lived, Hooke might have become the giant of his age. Had he not been endowed with his infamous, tetchy and quarrelsome nature and Newton himself not been so sensitive to criticism, both may have co-operated to make a great joint discovery. As it was, Hooke earned himself the title with his peers as the "irascible Dr Hooke". His reputation as a disputatious man followed him wherever he set foot. When Hooke was appointed surveyor and alongside Wren helped rebuild London after the Fire, he was best known to workmen as "Old Crotchety".

At meetings of the Society, and when not demonstrating, he was unable to sit still and to others round him he seemed in constant agitation. No portrait of him survives, and his unfortunate appearance relies on contemporary descriptions. He was born with a congenital defect so that his back was permanently bent. In stature he was even shorter than Wren. As he grew older, he became more round-shouldered and misshapen, and his deformed body gave him the look of a half-grown child. Invariably his long dark-brown hair remained uncombed. To his colleagues he seemed to spend hours stooped over his experiments and it is said he positively hated anything that resembled exercise. This was understandable, for physical activity would painfully remind him of his infirmity. Nevertheless, his diary confirms he rarely took a carriage and preferred walking along London's narrow streets, so this must have kept him in reasonable condition.

Hooke was inordinately vain about the powers of his intellect and his skill with his hands. As a model-maker there were no Newtons to challenge him. His mechanical expertise amazed colleagues and lowly artizans alike, who never equated this kind of ability with the mind and disposition of a scholar. John Evelyn said of Hooke: "No such person is to be found elsewhere in Europe for clever parts & ingenuity." Evelyn also noted astutely that Hooke had an uncanny and genuine ability to anticipate other people's discoveries.

When Hooke corresponded with Newton in 1679-80, the pair were drawn into a discussion about the path a body would follow if it could fall freely below the Earth's surface without it encountering any obstruction. This was the occasion when Newton committed a careless error in his reply. After Hooke capitalized on this, Newton allowed the correspondence to peter out, not wishing to embroil himself further in controversial matters which distracted him. Now, four years on, he began to write his book. When he started at the end of 1684, Newton, by his own admission, had no real conception where his pursuit would lead him.

Comets were certainly to figure prominently in his comprehensive treatment. In 1682 Newton had contacted Halley about his observations of comets, which included the "second" comet of 1680, made while the young astronomer was on the road to Paris at the start of his Grand Tour, for these were some of the earliest (as the comet re-emerged from near the Sun).

The exact date when Newton and Halley first corresponded and then later where they first met is not known. Newton had also solicited observations of the comets from all those who had likely made any. He soon discovered, however, that except for those of Flamsteed's and Halley's, the rest were too inexact to use. Newton wrote two long letters to Flamsteed about his attempted method in determining the paths of the comets using the basis of four positional observations, and three separate manuscripts survive from this attempt. At this time Newton held the view, like Kepler and others, that all comets moved in rectilinear paths, and while it was Flamsteed who maintained the two comets seen in 1680-1 were the same one, Newton still was of contrary opinion—rejecting Flamsteed's magnetic ideas that the Sun first attracted one pole of the comet, to explain its inward path, and then repelled the other pole, to explain its outward path. In rejecting Flamsteed's theory in toto, Newton may rightly be accused of throwing out the baby with the bathwater.

Newton was also distrustful of Flamsteed's dates for his observations of the comets. This was because the Astronomer Royal had confused New Style with Old Style dates. The confusion made nonsense of Newton's computed orbits which later had to be corrected. England had not yet adopted the Gregorian calendar, for this was considered a popish superstition, so that dates in England before 1700 were ten days out of phase with those of the Continent. Consequent to 1700, which England observed as a leap year, it was then eleven days out of phase after 28 February 1700. Manuscripts and letters in Halley's time often have a date notation 1/11 March, etc. In England, to compound the confusion, the new year legally began on 25 March. While some adhered strictly to the legal practice, others wrote double years (e.g. 1671/2) during the period from 1 January to 25 March.

In about 1679, Newton had already solved the mechanics of the motion of a planet orbiting the Sun, but he had not yet then attempted to extend the same principles to a comet. What held him back was the same lingering belief held by Kepler that comets were probably governed by a different set of God's laws. In Kepler's case another factor entering his thinking was that comets were probably unique bodies in a divine sense, and a comet might be sent by God as a sign. While Newton himself held strong religious beliefs, these did not extend to the use of comets as a divine tool. Contemporary feeling was that, physically, comets were anomalous foreign bodies not strictly related

30 Greenwich Park Observatory round Flamsteed's and Halley's time.

31 John Flamsteed (1646–1719)
er a portrait by Thomas Gibson.

to the rest of the solar system, and the opinion of Seneca, 1,600 years earlier, was still rejected. Even Hooke in his own writings on comets excluded them from the attractions between bodies, and as late as 1680 it is known that Halley held a similar view. Since 1665, Hooke must have changed his mind, for when lecturing about the Plague comet of 1664, he claimed it was a reappearance of the famous 1618 comet—with the inference it was a periodical body akin to the planets.

Consequent to 1680, Newton himself was having a rethink. In a surviving draft of a letter (Newton obsessively drafted everything several times before he committed it to print), it shows he did try to apply the dynamical qualities of planetary orbits to the 1680-1 comet(s). In spite of refuting Flamsteed's ideas, this was initially in terms of a magnetic attraction. In the final draft, which became the letter, he cut this passage and confined himself to a criticism of Flamsteed's (correct) assumption of one comet rather than two.

It was only after the appearance of the new comet in 1682 (Halley's comet) Newton began to doubt, then outrightly reject, the old Keplerian idea of the theory of rectilinear paths for comets. He now assumed that comet orbits must be curved ones and the point of greatest curvature would occur when the comet was closest to the Sun (at perihelion).

While Newton was unaware of Sir Thomas Lower's ideas on the subject back in 1610 (*see* p 176), he must have recalled that Seth Ward had put forward the idea of a circular or elliptical orbit for a comet seen in 1652, and Hevelius himself in his *Cometographia* in 1668 had vaguely suggested parabolic orbits for comets but without substantiating the idea. Newton, however, did not know that George Samuel Doerfell, an obscure Protestant clergyman from Saxony and a pupil of Hevelius, had already published a small book in 1680 consisting of five leaves in quarto and one woodcut showing that comets must move in parabolic orbits with the Sun occupying the focus.

CHAPTER 7

More Celestial Mechanics ... and Controversy

"Think of the myriad enmities, suspicions, animosities and conflicts that are now vanished with the dust and ashes of the men who knew them; and fret no more."—MARCUS AURELIUS

It was Halley's claim that in respect to Newton's *Principia* he had been the Ulysses who produced this Achilles. Yet this is only partially true. Much of the *Principia*, if not in manuscript, had long gestated in Newton's mind. This included the dynamics of the Earth-Moon system, the nature of the forces at work between the Sun and the planets and his schema, in outline, to bring to heel the erratic comets. But Halley certainly motivated Newton sufficiently for him to commit himself to write his brief, nine-page *De motu*.

Nevertheless, about the time Halley was submitting to the Society "a curious treatise *De motu*" and securing Newton's "invention", Newton seems to have been one step ahead of Halley and already planning the longer work. Proof of this is that after composing *De motu* he began to correspond with Flamsteed for new observations to enable him to make his demonstrations more precise. He told Flamsteed: "Now I am upon this subject I would gladly know ye bottom of it before I publish my papers."

In getting to the bottom of it Newton practically removed himself from Cambridge life. His behaviour became more erratic. So absorbed was his mind he often forgot to eat and, when outdoors, people noticed he sometimes took wrong directions only to stop abruptly. Then he would realize where he was supposed to be going and hurry away in the opposite direction. Once embarked on the project he knew his book could be the greatest intellectual achievement since Copernicus. Even so, as he progressed, doubts about his abilities to bring it off occasionally nagged him.

At the Royal Society the majority of Fellows found *De motu* way above their heads, even Flamsteed had not been able to understand it when shown it by Halley. The short book contained four theorems and five problems that considered motion in space void of resistance. With the more expansive work he planned, his mind began to reach out towards those horizons of thought so far only glimpsed from a distance. After reviewing his first definition and hypothesis to take in resistance, he added two more problems.

Much of the work was sheer drudgery. Some of his enquiries to Flamsteed concerned the celestial coordinates of two stars in Perseus near which the comet of 1680-1 had passed. The Plague comet he had observed as an undergraduate also interested him. He explained to Flamsteed: "I do intend to determin ye lines described by ye Comets of 1664 & 1680 according to ye principles of motion observed by ye Planets..."

Had Newton decided to include a comprehensive study of the comet of 1682, he might well have deprived Halley of his later prize in discovering its true periodicity, and today we would be calling it Newton's comet.

In *De motu* one of Newton's definitions drew upon a lesson in circular motion that Hooke had taught him during their correspondence in 1679. In it Newton now introduced a new word into the accepted vocabulary of mechanics: "I call that by which a body is impelled or attracted toward some point which is regarded as a centre *centripetal* force." He afterwards explained he had coined the word "centripetal"—seeking the centre—in a deliberate parallel to the word *centrifugal*—fleeing the centre—invented by Huygens not long before. It was the study of centripetal forces, as they determine orbital motion, which was to be a characteristic of Newton's work embodied by the *Principia*.

Parts of the original *De motu* that Newton sent to Halley underwent several revisions. Newton was one of those writers who choose to develop ideas by the method of successive drafts. Each one contained steps forward in his thinking over the previous. Soon he was corresponding with Halley, writing: "gravity is one kind of centripetal force: my calculations reveal that the centripetal force by which our Moon is held in her monthly orbit round the Earth is to the force of gravity at the surface of the

32 Newton's room (to the right of the Gate House) at Trinity College. It was here that Halley persuaded Newton to write the *Principia* and where Newton toiled for 18 months to produce his manuscript.

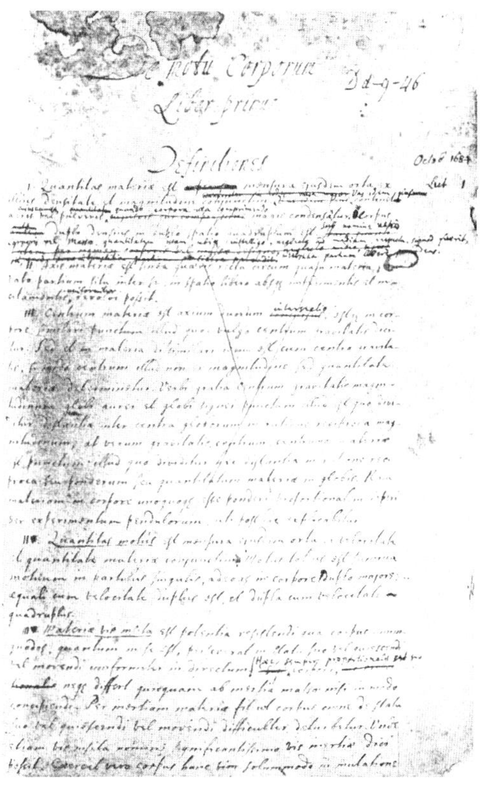
De motu Corporum
Liber primus
Definitiones
Octob 1684

When Newton embarked on his *Principia,* he also utilized his *De motu* notes for his Cambridge lectures in 1684.

Earth very nearly as the reciprocal of the square of the distance from the centre of the Earth."

Newton was making general headway with elaborating his main principles which he would assemble in the finished works as parts, or Books. Nevertheless, in a progress report to Halley, dated 20 June 1686, he confided that his theory of comets still remained unfinished. "In Autumn last I spent two months in calculations to no purpose for want of a good method, wch made me afterwards return to ye first Book & enlarge it wth divers Propositions some relating to comets."

It was in Book II he set out that comets definitely come under the laws of orbital motion he had formulated for planets, concluding: "... Therefore, during the whole time of their appearance, comets fall within the sphere of activity of the circumsolar force, and are acted upon by its impulse & these describe conic sections that have their foci in the centre of the Sun, and by radii drawn to the Sun describe areas proportional to the times as per Kepler's laws. For that force, propagated to an immense distance, will govern the motions of bodies far beyond the orbit of Saturn [then the farthest known planet]."

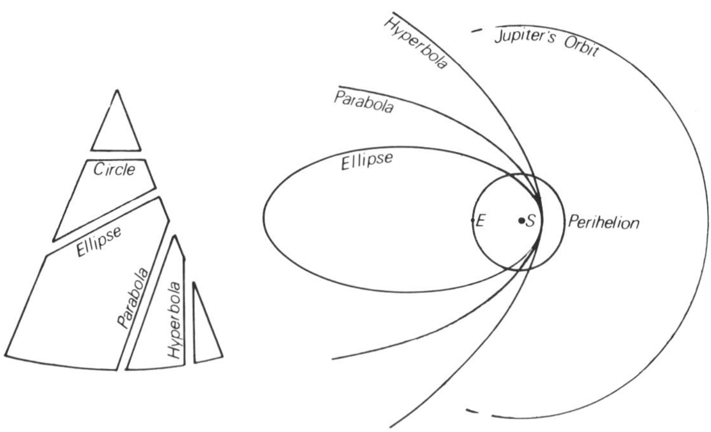

34 Conic sections and cometary orbits. (*left*) A cone can be sectioned to produce a circle, an ellipse, a parabola and a hyperbola. (*right*) Near the time of perihelion the different possible orbits for a comet are often indistinguishable from one another.

While descriptive theory is one thing, practical demonstration is quite another as Hooke had long realized. Newton was vexed he was unable to work out a practical method to put his ideas about comets into practice and determine an orbit from observatıons as one could do with the planets. The truth is he never did quite succeed in applying his method successfully to any particular comet. This was left to Halley in the years following.

During the course of his composition, Newton submitted his successive drafts to Halley, requesting his criticism. By now Newton had recognized that Halley was one of the few men in Europe to whom he could turn in absolute confidence and seek advice without involving disputation. Flamsteed, his other correspondent, was useless and like Hooke possessed only a smattering of mathematics. Newton was working beyond the frontiers of established knowledge and needed someone of comparable intellect as a sounding-board; someone sympathetic to the problems of a scholar in isolation.

It also needed someone of Halley's infinite tact to tread carefully through this potential minefield. Halley knew how ultrasensitive Newton was to criticism. He recognized that one false step, and the *Principia* might be irretrievably lost to mankind because of a sudden fit of pique. At the same time Halley was not always a fawning acolyte as his correspondence occasionally suggests; he was not afraid of offering constructive amendments, and some of these Newton adopted. Because of Newton's habit of working with successive drafts, and his unwillingness ever to throw away any scrap of manuscript concerned with his own thoughts, these amendments survive among his papers.

By 21 April 1686, Halley was writing an editor's puff to whet the fellowship's appetite. He announced the forthcoming "incomparable *Treatise of Motion*" was "almost ready for the Press" and this related "to all matter in the Universe". Halley was so caught up in lending assistance to Newton in 1686-7 that he sacrificed all his own spare time, and this was in addition to falling behind with the Society's correspondence.

It must have been sweet music to Newton's ears when Halley in his replies to letters waxed lyrical and referred to "your divine Treatise", adding: "Nearer the gods no mortal may approach." In his role of Clerk to the Society, Halley was unofficially Newton's editor. As his friend he was adviser and publicist and was constantly spreading flattering gossip among the fellowship at large about the impending masterpiece.

In April 1686, Book I was delivered to the Royal Society, and Halley made his official report to the Council. On 19 May the Society ordered "Mr Newton's *Philosophiae naturalis principia mathematica* to be printed forthwith in quarto and in a fair letter ... that a letter be written to him [Newton] to signify the society's resolution and to desire his opinion to the print, volume, cuts, & etc".

From the start Halley had been the driving force behind Newton's book. As Clerk to the Society he had constantly used his influence to promote the project. Now suddenly the financial responsibility of its printing also fell upon his shoulders. The circumstances of this are unclear, and where he found the money for this undertaking is not known. It was probably his wife's money rather than any remaining patrimony. There is no record of him ever asking Newton for a contribution towards its cost, and the Society itself was constantly short of funds. As Clerk he had not yet been paid his £50 stipend. Sitting in the President's chair at this time was Samuel Pepys. Since joining the Society in early 1665 he had subsequently lapsed as a Fellow for he considered much of its business above his head. It was his good friend Evelyn who begged him to come back in 1680 when the Society was in low water and needed the support of eminent public figures and experienced men of affairs.

Evelyn's flattery worked. The Society did pick up and two years later was making new rules for stricter consideration of candidates suitable for membership. Pepys was made President in 1686 and was still in office when the manuscript for the *Principia* arrived.

Shortly before, the Society had published Francis Willughby's *De Historia Piscium (The History of Fishes)*. It was a work dedicated to Pepys for reason he had dipped into his own pocket to pay for 79 of the 187 plates the book contained. The rest of the cost was shared by seventy Fellows. Five hundred copies were printed and sold at £1 5s on ordinary paper and £1 8s on best paper. The book was a commercial disaster and there were still unsold copies about as late as 1743. With fingers burned on this previous undertaking, the Society would not risk underwriting Newton's book—a work considered by most of the fellowship to be of an abstruse nature and which they suspected would enjoy a very limited audience.

Halley's decision to underwrite the printing certainly saved it from oblivion as the Society's financial affairs did not improve. Not only did Halley's £50 annual stipend remain unpaid, but in lieu of it, it was voted to give him fifty copies of the unsold said *Fishes* as compensation. Halley's reaction to this magnanimous decision by Council is unknown. He was probably too discreet to make a fuss about it. Likely he took his fifty copies and disposed of them for cash to Samuel Smith, a well-known printer-cum-remainder merchant of his day, who ran an establishment in

35 Diarist Samuel Pepys (1633–1703). Although not a virtuoso himself, he became involved in the affairs of the Royal Society for many years and was its President in 1686 when Newton's manuscript of the *Principia* arrived.

St Paul's Churchyard. In the following November, 1688, while William of Orange was quietly invading England, the Council of the Society, most likely on Halley's recommendation, ordered him to see if he could arrange further bargains with the same Sam Smith for another considerable parcel of "the Book de Piscibus".

After the *Principia* was printed and published with Pepys's imprimatur to the title page, it is doubtful if he gave it more than a cursory glance. It was all above his head, and he only vaguely comprehended what the purpose of the book was about. In a letter to his old tutor Joseph Hill, dated August 1688, he refers mischievously to two prevailing mysteries: one being "Mr Newton's *Principia* which is understood by few but deservedly admired by all those who understand it", and the second being "the activity of the Dutch Fleet" whose purport, Pepys hinted, seemed far more intelligent to the lay mind.

It was after the delivery of Book I of the *Principia* that a dark cloud appeared on Newton's horizon in the shape of a letter from Halley. It was the same letter in which he informed Newton he was now officially in charge of publication. After discussing some suggested changes in the diagrams, Halley dropped his other news as tactfully as possible. He explained he was personally disturbed on Newton's behalf, as his friend and confidant, by what he found necessary to pass on: "... I ought to inform you of, viz that Mr Hooke has some pretensions upon the invention of ye rule of the decrease of Gravity, being reciprocally as the squares of the distances from the centre. He sais you had the notion from him, though he ownes the Demonstration of the Curves generated thereby to be wholly your own; how much of this is so, you know best, as likewise what you have to do in this matter, only Mr Hook seems to expect you should make some mention of him, in the preface, which is possible, you may see reason to praefix. I must beg your pardon that is I, that send you this account, but I thought it my duty to let you know it, that so you may act accordingly; being in myself fully satisfied, that nothing but the greatest Candour imaginable, is to be expected from a person, who of all men has the least need to borrow reputation."

Newton replied immediately to Halley in an uncharacteristic mild tone that disguised his true feelings and misled Halley into thinking the author was not unduly concerned by this turn of events: "I thank you for wt you write concerning Mr Hook, for I desire that a good undertaking may be kept between us..." Newton then referred to events of 1679 and also requested Halley to speak to Wren apropos a conversation between them in 1677. Newton believed this would clarify for Halley what had gone on before he had joined the Society and show the nature of the uneasy relationship between himself and Hooke and how false was Hooke's claim.

In Book I, Newton had not mentioned Hooke's name once. His reason was he genuinely felt he owed nothing contained in this section to him. However, in another part of the work Newton had not yet sent to Halley, he had cited Hooke's name several times.

Following the dispatch of his reply to Halley, Hooke's insinuation began to prey on Newton's mind. He had been angered by Halley's news from London but had decided not to show it openly; nevertheless, as the days and weeks passed, he could hardly contain himself. The passage of time afforded no balm to his wounded pride but only seemed to inflame it. It took three weeks for Newton to express his true thoughts to his editor and friend in London. By this time Halley believed the worst was over. Newton's initial response, in view of Hooke's accusation, had been mild. Now Halley received another letter from Cambridge: "In order I let you know ye case between Mr Hook & me, I gave you an account of wt past between us in our letters so far as I could remember... I intended in this letter to let you understand ye case fully but it being a frivilous business, I shal content my self to give you ye heads of it short..." Then followed his detailed justification for his beliefs in the matter in relation to correspondence he and Hooke had exchanged in earlier years.

On the plain face of it there was justification for *some* kind of acknowledgement to Hooke, but Newton was reluctant ever to admit it.

Hooke, however, had only guessed at the inverse square law whereas Newton had actually demonstrated its mathematical truth. Had Newton possessed a more bending, forgiving nature, like Halley, he could have made the gesture and credited Hooke with part of the idea without any loss of kudos. Indeed, such an action might have enhanced his own standing with his peers at the Royal Society. Yet, Hooke only had himself to blame. Newton by now abominated his rival with a fanaticism which matched that he applied to his theological beliefs in Arianism. He could not and would not forgive Hooke's earlier transgressions: he would not yield an inch. After informing Halley Book II was to be a short one, and now only required copying, he explained that Book III, already written in draft, would be suppressed, for reasons "Philosophy is such an impertinently litigious Lady that a man had as good be engaged in Law suits as have to do with her. I found so formerly & now I no sooner come near her again".

As was his custom, Newton composed several drafts of this letter which survive for scrutiny. In the intervals between drafts he received news from an independent source in London that Hooke was bent on making trouble. It was now, in a postscript to the final draft he sent to Halley, he went as far as to accuse Hooke of publishing Borelli's idea under his name—commenting, sneeringly, that Hooke was incapable of making the necessary calculations to prove it.

In one draft of the *Principia*, Newton had started out to acknowledge Hooke's ideas about attraction. Now he blue-pencilled this from the final copy. In discussions about comets he had also referred to an observation made by "the very distinguished Hooke". This was now edited and shortened to a more simple "Hooke". At a later date the entire passage was eliminated, plus another in which he had cited Hooke's name.

What Halley had dreaded was now a reality. As editor and publisher he was in a dilemma. He was faced with an author bent on castrating his masterwork owing to a fit of pique. In a rejoinder to Newton he first flattered and then cajoled him—assuring him that the fellowship stood firmly behind him. On no account must he suppress Book III. He was recognized as the inventor of the inverse square law, and the whole Society was delighted by the honour he planned on paying them in his dedicaton of "so worthy a Treatise". He implored Newton: "Do not let your resentments run so high, as to deprive us of your third book, wherein the application of your Mathematical doctrine to the

Theory of Comets, and severall curious Experiments, which, as I guess by what you write, ought to compose it, will undoubtedly render it acceptable to those that will call themselves philosphers without Mathematics, which are by much the greater number."

In making his last point, Halley hinted broadly it would be unwise to alienate the majority element of the fellowship who were not versed in mathematics—exampled by President Samuel Pepys. Halley's letter apparently acted as the balm Newton sought. In his reply, Newton even went so far as to admit his debt to Hooke in three things. Nevertheless, he again flatly denied Hooke's accusation of plagiary. He was adamant: with no help or influence from others he had found the inverse-square law as his own, original, discovery.

Thereafter Hooke and Newton remained sworn enemies. Until his death in 1703, Hooke, in turn, never forgot nor forgave. He imagined Newton had robbed him of a glorious intellectual treasure; it constantly preyed on his mind and further blighted his personality. Newton was the intruder into the kingdom where Hooke once reigned unchallenged. Over the years Hooke had tried to belittle the achievements of the newcomer, but he had grown to fear him more than any other man. The irascible Hooke became even more irascible. Hooke's physical repulsiveness and his general demeanour in pressing his claim were against him from the start. It was unfortunately an example of Georg Lichtenberg's famous aphorism: some actions are called malicious because they are done by ugly people.

Hooke, however, did have his later champions. Ironically, the eighteenth-century savant Alexis Claude Clairaut, who was to become deeply involved during 1757-8 in applying the principles of Newtonian universal gravitation to the orbit of Halley's comet, was always a staunch supporter of Hooke's claim.

Halley, as Newton's editor, friend and adviser, had been his most vociferous champion. As may be expected this did not endear him to the man who had proposed him for fellowship of the Society back in 1678. Hooke had a reputation for handing out retribution as well as reward. It seemed to him that Halley had now bitten the hand that once had fed him.

Meantime, Newton had agreed not to suppress Book III.

Originally, as Halley had reminded him in one of his letters, it was Newton's intentions to use this as an opportunity for a prose essay—a leavening to his majority non-mathematical audience. It was important, as Halley had pointed out, not to make the subject-matter too difficult to grasp. Newton now had a change of heart: his desire to revenge Hooke's slight on his character required expiation. He informed Halley that he would show the "Smatterer in Mathematics", as he labelled Hooke, what his theories were all about. Sweet revenge for Newton was to publish the material in a form Hooke would be unable to follow.

From hereon Hooke's star waned. History has generally been unkind to his particular genius as a polymath. Even in the field of architecture he is overshadowed by Wren, who unjustly has taken credit for several of Hooke's brilliant designs. He was his own worst enemy; he antagonized all but his closest friends and at times even tried their patience. Although he remained a bachelor, like Newton, and outwardly was unattractive to the opposite sex, he was a man possessed with strong sexual passions. Whereas Newton, from all the signs, was undoubtedly a passive homosexual (there is little evidence to suppose he was an active one), Hooke was a proven womanizer. At Gresham's, where he was Professor of Geometry, he conducted torrid affairs with successive housekeepers. Like his contemporary Pepys he entered details of his dalliances in his diary—recording his successful orgasms cryptically, choosing the zodiacal sign of Pisces because, it is said, of its association of Venus and Cupid in one of the origins of this symbol. More likely, however, it was because the two fishes are usually shown entwined. In his diary there are also admissions of other sexual scandals. When his favourite niece, Grace, the daughter of a brother who committed suicide, replaced a housekeeper, he fell victim to the grand passion of his life. Now middle-aged, he doted on her, but her unfaithfulness drove him to distraction. The year 1687 was a black one for him: first the publication of the *Principia* and the general accolade accorded his rival, then the death of his niece. Post 1687, Hooke's decline as a man and a scientist was a rapid one.

Newton's revised Book III contained an important addition. He believed he had cracked the problem of the comets and they had finally yielded to his mathematical assault. Later he commented: "this discussion about comets is the most difficult in the whole book."

Newton now claimed that in theory at least it was possible to compute a comet's orbit. From three evenly spaced observations, measured against its shift across the background stars, a body such as a comet moving in a different plane to that of the planets could be pinned down with exactitude. Applied to a specific comet, however, the method depended on placing the middle observation on the orbit by an informed guess. If the first guess were incorrect, it could, nevertheless, be corrected by a repeat process of hit or miss until the observation fitted.

Using this trial and error graphical method for the comet of 1680-1, he found it moved in a plane inclined to the path of the major planets (the ecliptic) at an angle tilted at $61°20\frac{1}{3}'$. This angle agrees fairly well with the more modern calculation of $60°40'$. By locating the point of the comet's perihelion, Newton was then able to fix the axis of its orbit.

Newton concluded that the 1680-1 comet travelled round the Sun in a parabolic orbit. He was wrong about this, but as a pioneer effort his demonstration was a pretty good attempt. Even if the derived curve was the wrong one, his thinking was sound. Some years later, Halley, after refining the method, found the 1680-1 comet moved in a closed ellipse and not in an open-ended parabola, and this gave it a periodicity of 575 years. Halley too was only partly correct. The shape of the comet's orbit was probably right but not its period.

Although Newton was wrong about his choice of a parabola, he had, nevertheless, proved conclusively that planetary laws governed the motion of the 1680-1 comet in its path round the Sun and, by inference, the motions of all known comets.

And to be strictly fair to Newton, by this time even he favoured elliptical orbits for comets but had insufficient data to prove this. His choice of a parabola for the 1680-1 comet was therefore a compromise solution for its orbit... Even today, for convenience, all new comet orbits are first computed by assuming parabolas.

CHAPTER 8

More Controversy...

During the winter of 1686-7, Newton had finished his manuscript, but its printing was delayed because of events affecting Halley's position as Clerk to the Society. Halley had not reckoned with Hooke's vindictive nature and his behind-the-scenes manoeuvrings. It was now his performance as Clerk was suddenly called into question.

On 29 November 1686 the Council of the Royal Society were persuaded to make a vote on Halley's continuation in office or whether a new clerk should be elected in his place. In Hooke's eyes, Halley as accomplice, was as guilty as Newton for "stealing" his ideas on action at distance.

Fortunately for Halley, while he had neglected his personal interests to help Newton, he had been reasonably diligent in the Society's day-to-day affairs—except, and there can be little doubt, correspondence was behind. His energy, his winning ways and his pleasing countenance were much admired by the majority of the fellowship. He had enemies, but what man in public office hasn't! Since his appointment he had injected new purpose into the weekly meetings, and the *Philosophical Transactions*, the official publication of the Society which before his appointment had been allowed to lapse, was started up again.

It had been significant that Hooke had been elected to the Council that year. His voice carried new influence, and he was a force to be reckoned with. During the winter there was lobbying by Hooke's cronies to oust the popular incumbent Clerk. Another factor was Hooke's undoubted jealousy of the success that Halley had made of the Clerkship. Before Halley's time Hooke, as curator of experiments, was responsible for "entertaining" fellows at the weekly meetings. Although there can be no question he was a scientist of the first order, as an entertainer his personality worked constantly against him—especially as he grew older and even more crotchety. Yet another factor which had brought on the crisis was that Halley and Hooke appear to have been in direct competition for use of the Society's limited funds.

On 5 January 1687 a committee was appointed to investigate Halley's conduct, and Hooke was invited to bring in a proposal to supply meetings, for a salary, "with experiments and discourse".

For a time things looked black for Halley's future, but his friends rallied to help him. In a letter to Newton, dated February 1687, explaining circumstances in London, he reported that "6 of 38, last generall Election day, did endeavour to have me put by". Halley maintained a brave face and refused to involve himself in vindictive counterploys. His detachment did much to impress the neutral members of the committee. When the committee met again and reported in his favour, he wrote to Newton the following day about immediate plans for pushing ahead with the publication of the *Principia*.

In the pre-typewriter age Halley spared Newton much of the drudgery in providing the printer with fair copy. He told a friend in a letter: "the corrections of the press costs me a great deal of time & paines." As reward Halley was later able to read a handsome acknowledgement from Newton in his preface: "In the publication of this work the most acute and universally learned *Mr Edmund Halley* not only assisted me in correcting the errors of the press and preparing the geometrical figures, but it was through his solicitations that it came to be published; for when he had obtained of me demonstrations of the figure of the celestial orbits, he continually pressed me to communicate the same to the Royal Society."

If Newton ever expressed his more private thanks by correspondence, this letter has been lost to posterity. But Newton was eternally grateful and in later years referred generously to the *Principia* as "Halley's book".

There were more delays due to repeated corrections to printers' errors because of the complexity of the text. At the start

Halley considered using three different printers—one for each book—but two sufficed. At last Halley was able to report to Newton on 5 July 1687: "I have at length brought your Book to an end, and hope it will please you. the last errata came just in time to be inserted. I present from you, the books you desire to the R. Society, Mr Boyle, Mr Pagit, Mr Flamsteed and if there be any elce in town that you design to gratifie that way; and I have sent you to bestow on your friends in the University 20 Copies, which I entreat you accept."

The finished work was a 500-page masterpiece containing 97 propositions in the first book, 53 in the second and 42 in the third. It was priced nine shillings. Halley had decided to produce it in small quarto and, of course, in Latin as was custom. The title page now read *Philosophiae naturalis principia mathematica*, and while the inscription was dated 5 July 1687, some copies were distributed earlier. How many copies were printed is unknown as is also Halley's profit or loss in the venture.

Sam Pepys was not the only one to be baffled by the *Principia*. Newton presented a copy to one of his colleagues at Trinity, a Dr Babington, who reported after receipt of it "he might study it seven years" before he understood anything of it.

Anecdotes in a similar vein are common of the period. One relates that just after publication, when Newton passed a student and his companion in a Cambridge street, the student was overheard to remark: 'There goes the man that writt a book that neither he nor any body else understands."

However, among the cognoscenti its reputation was soon awesome and this spilled over to the level of the common man. Booksellers were inundated by a curious, non-Latin-reader public who wished to buy "a translation" of it; but apparently "translation" did not only refer to the language problem.

Not content with what he had done already to publicize the work, Halley composed an anonymous puff for it in the *Philosophical Transactions*. It was couched in a rhetorical phraseology that might bring a blush even to the faces of twentieth-century exponents of the publishers' art: "This incomparable Author having at length been persuaded upon to appear in publick has in this Treatise given a most notable instance of the extent of the

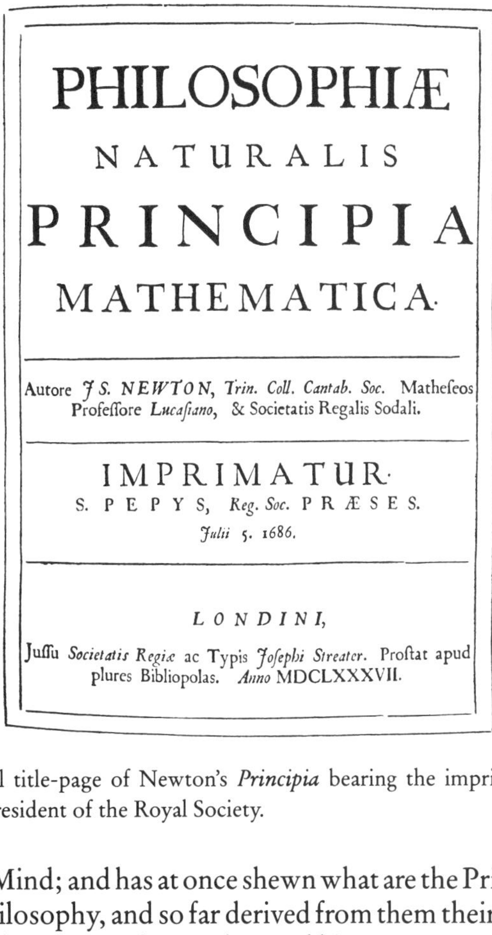

PHILOSOPHIÆ
NATURALIS
PRINCIPIA
MATHEMATICA.

Autore *J S. NEWTON, Trin. Coll. Cantab. Soc.* Mathefeos Profeffore *Lucafiano*, & Societatis Regalis Sodali.

IMPRIMATUR.
S. PEPYS, *Reg. Soc.* PRÆSES.
Julii 5. 1686.

LONDINI,

Juffu *Societatis Regiæ* ac Typis *Jofephi Streater*. Proftat apud plures Bibliopolas. *Anno* MDCLXXXVII.

36 The original title-page of Newton's *Principia* bearing the imprimatur of Samuel Pepys, President of the Royal Society.

power of the Mind; and has at once shewn what are the Principles of Natural Philosophy, and so far derived from them their consequences, that he seems to have exhausted his Argument, and left little to be done by those that shall succeed him..." And as the man better equipped than most of his peers to appreciate the magnitude and sheer genius incorporated in the *Principia*, Halley then added: "... it may be justly said that so many and so Valuable *Philosophical Truths* as are herein discovered and put past Dispute, were never yet owing to the Capacity and Industry of any one Man."

Halley's panegyrics spilled over into the printed pages of the *Principia* with an eulogy to Newton, expressed in some high-flown Latin hexameters. However, a factor which needs to be remembered in the context of Halley's puffs and promotions is he was not only a friend and confidant but also editor and publisher.

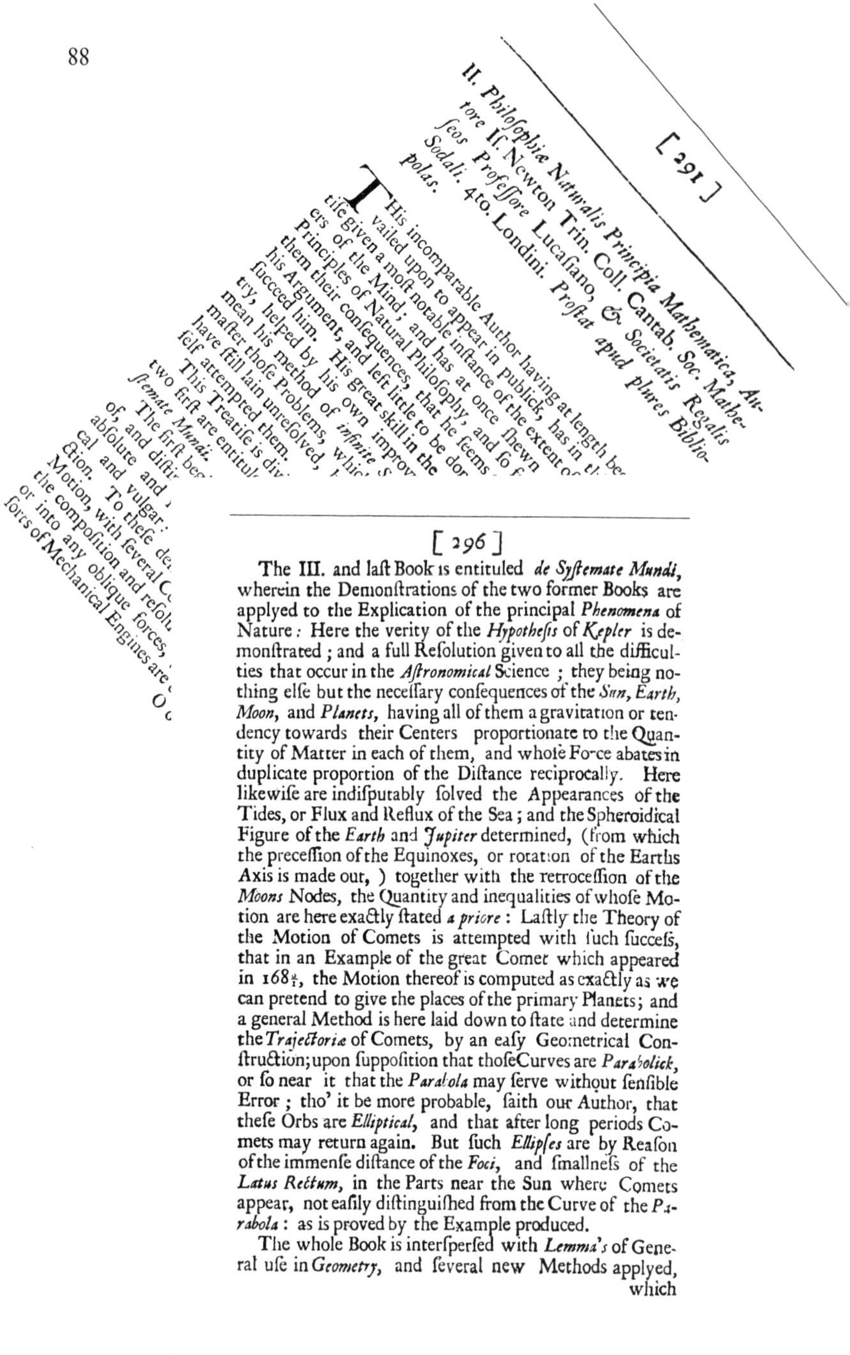

[296]

The III. and last Book is entituled *de Systemate Mundi*, wherein the Demonstrations of the two former Books are applyed to the Explication of the principal *Phenomena* of Nature: Here the verity of the *Hypothesis* of *Kepler* is demonstrated; and a full Resolution given to all the difficulties that occur in the *Astronomical* Science; they being nothing else but the necessary consequences of the *Sun, Earth, Moon*, and *Planets*, having all of them a gravitation or tendency towards their Centers proportionate to the Quantity of Matter in each of them, and whose Force abates in duplicate proportion of the Distance reciprocally. Here likewise are indisputably solved the Appearances of the Tides, or Flux and Reflux of the Sea; and the Spheroidical Figure of the *Earth* and *Jupiter* determined, (from which the precession of the Equinoxes, or rotation of the Earths Axis is made out,) together with the retrocession of the *Moons* Nodes, the Quantity and inequalities of whose Motion are here exactly stated *a priore*: Lastly the Theory of the Motion of Comets is attempted with such success, that in an Example of the great Comet which appeared in 168⁴⁄₁, the Motion thereof is computed as exactly as we can pretend to give the places of the primary Planets; and a general Method is here laid down to state and determine the *Trajectoria* of Comets, by an easy Geometrical Construction; upon supposition that those Curves are *Parabolick*, or so near it that the *Parabola* may serve without sensible Error; tho' it be more probable, saith our Author, that these Orbs are *Elliptical*, and that after long periods Comets may return again. But such *Ellipses* are by Reason of the immense distance of the *Foci*, and smallness of the *Latus Rectum*, in the Parts near the Sun where Comets appear, not easily distinguished from the Curve of the *Parabola*: as is proved by the Example produced.

The whole Book is interspersed with *Lemma's* of General use in *Geometry*, and several new Methods applyed, which

37 Two folios from Halley's puff which appeared in the *Philosophical Transactions* to launch Newton's *Principia*.

While there can be no doubt he genuinely felt his expressed admiration for Newton's genius, this must have been tempered with the desire, unconscious or otherwise, to get his money back, if not make a profit.

When the *Principia* took Britain by storm, Halley's expectations were fully realized. From hereon Newton's reputation was assured; he was recognized as a scientific giant among his contemporaries. The publication of the *Principia*, a revolution in thinking, preceded the political revolution which now threatened. In November 1688, William of Orange landed his army; by Christmas the same year, James II had fled the country for good.

Under the new order, Newton, the one-time recluse and introverted don, immersed himself in politics and aligned himself with the more extreme Whig faction. Earlier in the decade he had surprised his colleagues at Cambridge with his vigorous defence of the rights of the University against the encroachments of James II, defying the notorious Judge Jeffreys to do his worst. This stood him in good stead as a champion in the new democracy. In the Convention Parliament of 1689-90 he secured himself a seat.

As for Halley, the Glorious Revolution had little immediate impact on his career. Free of the drudgery of seeing Newton's *Principia* through the press, he busied himself with new scientific papers. In the years following the publication of Newton's book, the pair had little contact but they did occasionally meet socially. There was no envy in Halley; he was only too pleased for Newton's success. Among his contemporaries he received scant recognition for the key part he had played in getting Newton to present his genius to the world. Only in later generations was Halley's vital role publicly recognized. Augustus De Morgan, one of his nineteenth-century champions, concluded: "But for him in all probability, the work would not have been thought of, nor when thought of written, nor when written printed." While this view perhaps overstates the case, it represents a near-truth.

Halley's domestic life during the years following the *Principia* is practically unknown. We know two daughters, Catherine and Margaret, were both born in 1688, but we have no information of what his wife, Mary, felt about his self-sacrifice and the long hours spent in the cause of another's book—nor her feelings about her husband's financial underwriting of it. Halley was now thirty-one and at the crossroads of his life; he seems to have recognized his earlier, meteoric, promise had been slowed down of late and that his Clerkship to the Society if it continued might limit his future. These immediate post-*Principia* years were probably frustrating ones for him, yet scientific papers flowed uninterrupted from his pen.

The year the *Principia* was published, Halley wrote three papers—the third of which was to influence the course of his life over the next few years. It was the resurrection of a theme earlier treated by Hooke and others: namely that the event of the biblical Flood had been caused by a gradual shift of the Earth's poles. Halley took up the bones of the idea and concluded the Flood must have occurred in the very distant past. He was treading on dangerous ground. Not because he had borrowed a theme from Hooke but because of the Church. The inference from Halley's paper was that the time-scale then accepted for the Creation was in error. While Halley did not directly and openly dispute the generally accepted figure, the inference was enough to alert the bishops to a possible new heretic now among their midst.

What Halley suggested was the Flood must have been caused by another factor. One cause that readily sprang to his mind was the possibility of the Earth's close encounter with a comet or another similar wandering cosmic body. In his arguments he cited the fact that pagan, as well as biblical, sources indicated the deluge must have been a sudden catastrophic event. A comet, therefore, would aptly supply the means to fit the event and bring about a gravitational upheaval on Earth. Although Newton's work had shown that comets were subject to universal gravitation, he had not yet shed light on the nature or on the masses of comets. Halley, nevertheless, believed that such a close encounter with a comet might change the length of the year and the shape (the eccentricity) of the Earth's orbit. Halley's ideas were to have a profound influence on William Whiston and on the pseudo-scientific theories of Immanuel Velikovsky and innumerable other modern catastrophists.

Interests in biblical chronology and antiquity were shared by both Halley and Newton. Newton published little in his lifetime, but Halley was less prudent. Theologists and good churchmen disapproved of "dabblers in biblical chronology". James Usher, the Irish Archbishop, had dated "the beginning" to exactly 9 am 23 October 4004 BC, and this date was then considered to be as reliable and sacrosanct as the gospels themselves. Halley's obliquely hinted contrary opinion was to have important repercussions.

In 1688 he was briefly on his travels again when he visited Holland on behalf of the Royal Society to see Huygens and other Dutch scientists. Very little is known about this episode except he left from Harwich and arrived home with a book about the 1680-1 comet written in High Dutch.

As the years ticked by, papers continued unabated. The year 1691 saw Halley publish six scientific papers on diverse topics. Science then was not subject to divisions and rigid water-tight disciplines, later introduced by the British Association for the Advancement of Science in Victorian times, which in the twentieth century resulted in the dichotomous "two cultures". Halley's second paper of that year is of particular interest as it represents a pioneer effort to date historical events by fixing them to contemporaneous astronomical events that could be calculated accurately. This paper was entitled: "A Discourse leading to prove at what time and place Julius Caesar made his first descent upon Britain." In it Halley showed that evidence for the exact time and date might be gleaned by a precise recalculation of the time of an eclipse, for an eclipse of the Moon was reported to have taken place at the time Caesar landed.

About this period Halley was also involved in the practical problems of deep-sea diving. In 1689 he had published a paper entitled "A Method of Walking under Water". Since his voyage to St Helena, the sea and all things nautical had constantly fascinated him. Round about the end of the 1680s, sea-diving as a new technology was much in vogue. It was realized that if men could descend to the bottom of the sea, vast treasures lost in the past might be recovered. No doubt at the back of his mind Halley relished the prospects of such expeditions to make his fortune. One such expedition, which had been the talk of London, is briefly referred to in Evelyn's diary in 1687 about "a Spanish Galioon" that had perished somewhere "near Hispaniola [the Bahamas]" and had been raised by divers. The spoils created a vast fortune for its backers as well as the King.

By June 1691 we hear Halley was temporarily residing at the small seaport of Pagham, in Sussex, engaged in experiments with a diving-bell, operating from a naval frigate loaned for the exercise. Ever the adventurer, Halley himself descended to the seabed and reported back his experiences to the Royal Society. It was also round this time he received his first serious rebuff to the further advancement of his career.

We can assume he had no wish to remain Clerk to the Society for the rest of his life. We can imagine he was constantly on the look-out for chances to improve his status. An opportunity arose when his old Oxford professor, Edward Bernard, resigned the Savilian Chair of Astronomy—a post Wren had once held. Halley was still busy in Pagham with his diving-experiments when the news reached him. He immediately entered his name for the vacancy, realizing he stood an excellent chance.

The appointment of the new professor had to be approved by the Church, and soon disturbing news reached him, making him aware that some prominent clerics had not forgotten his "tampering with chronology" and were openly speaking out against him.

In a letter sent from Pagham to his influential friend Abraham Hill, Treasurer of the Royal Society, Halley writes: "an affair of great consequence to myself call me to London, viz Looking after the Astronomy-Professor's place in Oxford. I humbly beg of you to intercede for me with the archbishop Dr Tillotson, to defer the election for some short time, 'till I have done here, if it be but for a fortnight: but it must be done with expedition, lest it be too late to speak. This time will give me an opportunity to clear myself in another matter ... there being a caveat entered against me, till I can shew that I am not guilty of asserting the eternity of the world."

Shades of Bruno and Galileo!

Whatever the case against Halley, other candidates with high reputations were vying for the vacancy. A new acolyte of

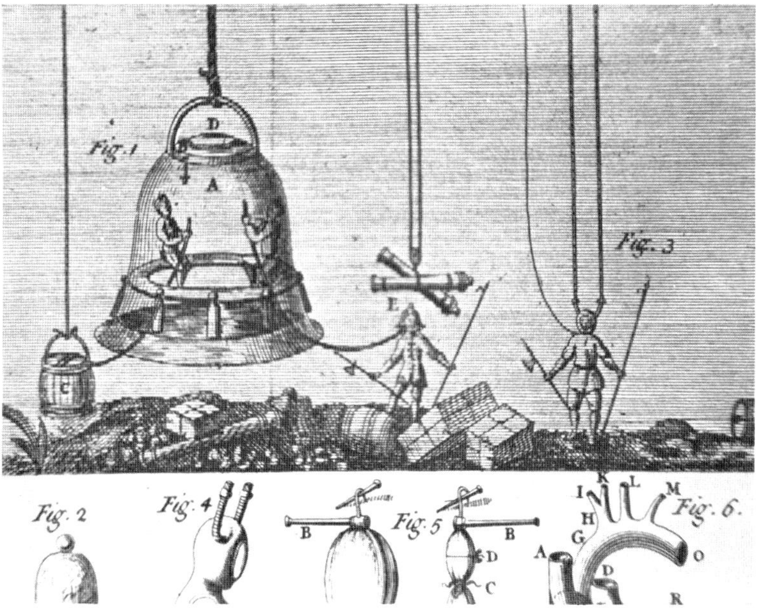

38 In the late 1680s and early 1690s Halley was engaged in underwater experiments with a diving-bell like the one depicted here.

Newton, David Gregory, was a front runner. Gregory, Professor of Mathematics at Edinburgh, had sought to establish a friendship with Newton by correspondence (since the *Principia*, striking up friendship with the Cambridge genius was a growth industry). David Gregory was nephew to James Gregory who had invented a design for a reflecting telescope before Newton. Like the Cassinis of Paris, the Gregory clan in Scotland was to spawn a succession of notable scientists.

From the start David Gregory's motives were his own self-interest. He courted Newton's favour in a shameless manner by extravagant flattery. Whereas Halley's praises respecting the *Principia* were genuine and from the heart, Gregory's were calculating. In letters he gushed with phrases like: "Farewell, noble sir, and proceed as you do to advance philosophy beyond the paths of Sun & sky." He must have had a liking for this particular turn of phrase, for, in correspondence with Huygens three years later, he dusts if off again.

One of the referees of the Church for the professorship appointment was Edward Stillingfleet, then Bishop of Worcester. He was a churchman well known for his orthodoxy and his vociferous opinion about the supreme authority of the scriptures. He was two years younger than Pepys, who had often heard him preach in London as a young man. Pepys believed him plain, honest and a good orator yet thought his book *Origines Sacrae* "frivolous in parts".

According to Thomas Hearne, not always the most reliable and unprejudiced chronicler of Halleiana gossip, the bishop, during his interview with the candidate, asked him some religious questions. Halley is reported to have retorted: "My Lord, that is not the business I came about. I declare myself a Christian and hope to be treated as such."

William Whiston, who succeeded Newton as Professor of Mathematics at Cambridge, relates that Halley at the interview was asked to recant his disbeliefs about the Deluge but stubbornly refused to do so on principle.

Halley was to be hounded by accusations of irreligious beliefs throughout his life. He is reported to have upset Newton over his non-orthodox views on several occasions, not to mention the over-pious Flamsteed. In truth, Halley's beliefs were based on little more than an honest scepticism which took the form of deism: God exists and indeed created the world but thereafter assumed no control over it or its people.

Halley must have arrived at the bishop's interview bucked by the knowledge that his old Oxford college was fully behind him. "We judge him to be in every way the most fit and accomplished for this performance as well as from our own long experience of his mathematical genius, probity, sobriety, and good life, as from the very many testimonials of all foreigners."

The Council of the Royal Society also gave him a recommendation "signifying their opinion of his abilities to perform the office of professor of astronomy now vacant, as likewise to testify what he had done for the advancement of the said science".

Before seeing Stillingfleet it is likely that the bishop's chaplain, Richard Bentley, was asked to look into the nature of Halley's alleged heresies so they could be laid before him for clarification.

Bentley is a curious character in the story of Halley and Newton. From a distance he had long admired Newton and was jealous of men like Halley with superior intellect who earned his respect. Later, after ingratiating himself into Newton's favour, he was to be involved with the second edition of the *Principia* when he made a point of not consulting Halley once over changes and tampered with the Latin verses, composed in praise of Newton, contained in the first edition.

Whatever occurred at the interview, Halley was not appointed. The chair went to David Gregory, who had obtained a recommendation from Newton. Why Newton championed Gregory rather than his old friend Halley—the man he owed more to than any other—can only be guessed at. There may have been a disagreement between them following the publication of the *Principia*. Perhaps over Halley's profits as publisher? However, all is conjecture. A more likely reason is that Newton knew of the Church's displeasure with Halley; perhaps he did not wish to be tainted by the same heresies by taking Halley's part. He was now making his way in public life and mixing in open society and it is well known he had a lifelong dislike of controversy. Supporting Gregory rather than his old friend might well have been his means of saving face (*see* note 6).

How Halley reacted to this betrayal by Newton is unknown. He was never one to bear grudges and he probably accepted it phlegmatically. He knew, like everyone else, he was Gregory's intellectual superior, but he refused to resent him personally. Proof of this is that in later years he and Gregory co-operated amicably on some translations of Greek and Arabic geometries.

Flamsteed had probably also played a role in putting a spoke in Halley's aspirations for the Oxford professorship. The pious Flamsteed had by now turned vehemently against his old student protégé for some alleged affront. Halley's lack of orthodox religion may have been one reason, at least Flamsteed always hinted at this. Perhaps the real truth, however, is that Halley is said to have caught Flamsteed out in some mathematical calculation. When Halley pointed out the error, the Astronomer Royal took umbrage and considered it a slight against his character. Yet another alleged factor was that Flamsteed had blatantly published some of Halley's tables as his own. Whatever the reasons the friendship deteriorated soon after Halley's return from St Helena. Flamsteed, in turn, went as far as to accuse Halley of buying the papers of Peter Perkins (the Mathematical Master of Christ's Hospital) who was a good friend of Halley, and then publishing some of them under his own name. This imagined plagiary was proved to be nonsense, but it was yet another factor that widened the split between them (*see* note 7).

To Newton, Flamsteed justified his attacks on Halley's character on the ground it was his Christian duty as a clergyman to save Newton from Halley's impiety. On the occasion of the vacancy at Oxford he wrote vindictively to Newton exhorting him to prevent Halley's election for reason he would "corrupt ye youth of ye University with his lewd discourse". On another occasion when he attacked Halley he counselled Newton to read the 9th and 10th chapters of Jeremiah about deceivers and liars.

The preferment of Gregory over Halley provided the gossips with some wry tavern humour. The best-known story is probably that of a certain Scotsman who, on hearing that Halley had come unstuck with the bishop, immediately packed his bag and travelled to London "to meet a man with less religion than Gregory"!

Halley appears not to have taken his set-back too seriously. He was still only thirty-five; there were many papers to write, and he was generally optimistic for the future. The sea, ships and navigational problems continued to fascinate him. During 1692 most of the subjects he tackled had a nautical slant. They ranged from "An instrument for Measuring the Way of a Ship" to an idea for the design of a quadrant with telescopic sights. Such an instrument had earlier been described by Hooke. For his pains in trying to update the idea, Halley was to be in receipt of a typical sour Hooke rejoinder about having done this himself previously. In another paper he described "A Method of Enabling a Ship to Carry Guns in Bad Weather", and this was an idea which Hooke sneeringly dismissed as "absurd".

On clear nights he pursued his astronomical observations, especially those of recording the Moon's position. The Moon's changing position against the stars was then held to be the key to mariners finding accurate longitude at sea; but first a more exact theory of the Moon's motion round the Earth was needed. Newton himself toiled long at this problem without ever solving it.

Another problem that fascinated Halley was the rate at which sea-water evaporated. Could this, he wondered, provide a clue to the puzzle about the age of the oceans and then in turn the age of the Earth? He argued that if the oceans had begun as fresh water, their increasing salinity as they dissolved more salt from rocks century by century and millennium by millennium would be a direct measure of time. From his studies he concluded a great age for the Earth, far in excess of Archbishop Ushers's estimate based on the Book of Genesis.

The Earth's prehistory fascinated him. Today, Halley would have revelled in the great geological debates on sea-floor spreading and continental drift as well as those of more recent antiquity concerning early man's uses of Megalithic sites as observatories—such as Stonehenge. He was the first to attach great antiquity to places like Stonehenge and Avebury—remarking that Stonehenge, from the appearance of its stones, must be at least 3,000 years old, whereas others dated them as Roman monuments. Comets as agencies for change wrought on Earth continued to attract him. He suggested the Caspian Sea was a water-filled depression caused by a direct impact with a comet. He even computed the volume of water of the resulting flood. Long before the age of Baron von Humboldt and Charles Darwin he surmised that beds of sea-shells found above present-day levels of high water indicated past great convulsions of the Earth's crust.

CHAPTER 9

Comets and Diversions

In the middle 1690s, comets were again at the forefront of Halley's thinking. Since the *Principia* he and Newton had little but the occasional social contact; then, in 1695, they re-entered into a vigorous correspondence about one of their mutual favourite topics. If Halley had borne any resentment over Newton's lack of support for the professorship at Oxford, he did not allow it to colour their relationship.

In the *Principia*, Newton had argued that the paths of comets near the Sun were curved like parabolas. In his researches into comets he had had to consider a choice of three possible curves: a closed one—the ellipse, as described by the planets; the open-ended parabola; or the hyperbola (Fig 34). Newton rejected the closed ellipse and the hyperbola. This was in spite of earlier ideas shared by several people that ellipses probably suited the paths of comets as well as they did the paths of planets. However, these were only speculations without mathematical proof. Newton had made the break-through and provided the geometry and then the calculus to demonstrate, by formal proof, that a parabola did suit the observations for the comet of 1680-1.

Even so, Newton would have preferred to have adopted the ellipse for all comet orbits but lacked the observational data to prove it. This had alerted Halley to the possibilities for future, more exacting, researches...

In a letter to Newton, dated 5 April 1687 when the *Principia* was being finalized in the press, Halley had written: "... Your method of determining the orb of a comet deserves to be practised upon more of them, as far as may ascertain whether any of those that have passed in former times may have returned again; for their nodes and perihelion being fixed, will prove it sufficiently, and by their periods, the transverse diameters will be given, which possibly may render the problem more easy..."

Indeed, Halley at this time had put his finger on those key elements necessary to understand how comets moved, but it took him several years before he now (in 1695) acted on it himself. In a later passage of the same (1687) letter, Halley, with a suggestion of reproof to Newton, comments: "... I do not find that you have touched that notable appearance of comets' tails , and their opposition to the Sun which seems rather to argue an efflux from the Sun rather than a gravitation towards it."

Yet Newton never carried his own researches about the physical nature of comets much further. Since Seneca's time, 1,600 years earlier, cometary astronomy generally had progressed little. The only notable advances before the *Principia* was the realization that comet tails for some unknown reason always pointed away from the Sun. In the West the earliest references to the recognition of this phenomenon is by Peter Apian (Petrus Apianus) in 1531, although if one reads Seneca carefully, there is already a hint of it there. In China, where comets were much observed by a bureau of civil servant "star-clerks", the first documentary evidence about the unique positioning of comet tails dates to AD 800, but this knowledge also probably refers back to much earlier times.

39 Comet tails are generally directed away from the Sun.

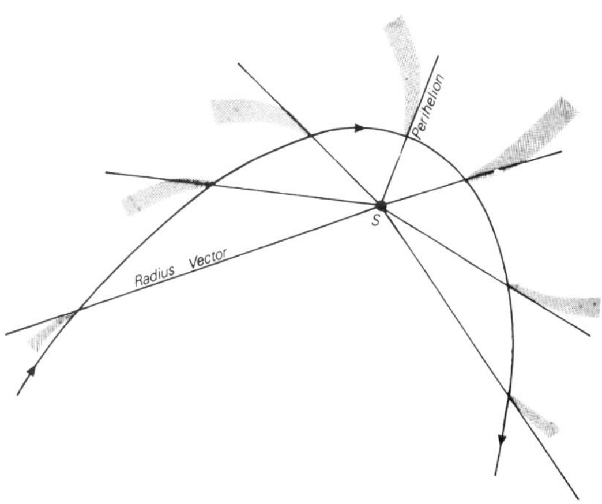

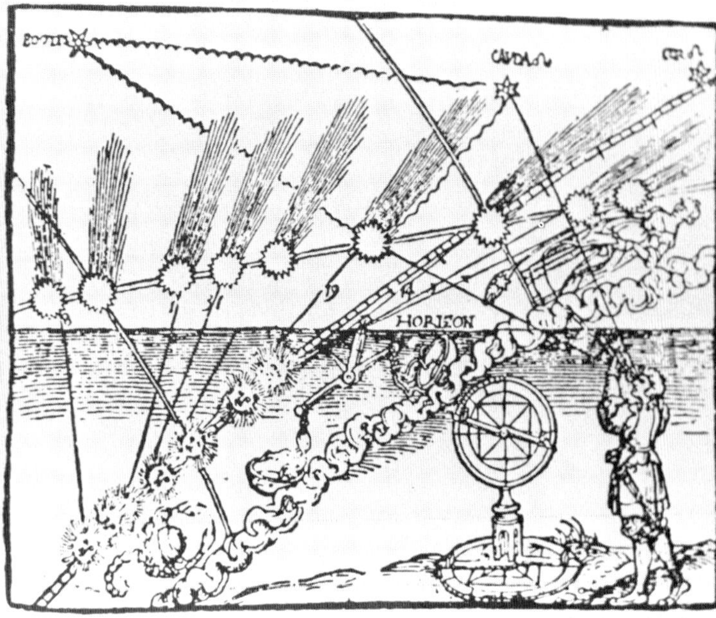

40 Peter Apian observed Halley's comet in 1531 and also the comet of 1532 whose path is depicted in this old German woodcut constructed to demonstrate the proof that comet tails are generally directed away from the Sun. This fact, however, was already known to the Chinese since AD 837.

In the fifteenth century, Johann Müller (Regiomontanus) had claimed comets were objects lying far beyond the Earth's atmosphere, and Cardan (Cardarnus), the following century, had much the same idea. It was not, however, until 1577, when Tycho Brahe actually proved the comet of that year lay beyond the Moon, that people took the idea seriously. Galileo, as late as the 1630s, still refused to admit comets were true astronomic bodies and he favoured Aristotle's outdated ideas. For this reason comets, if they were studied at all at universities, were included as part of *physica*, along with other atmospheric phenomena.

Kepler had been much interested in comets from the time as a six-year-old his mother carried him from his bed to see the spectacle of the 1577 comet, and it was this event apparently that kindled his lifelong interest in astronomy. Subsequently he became curious about comet tails and professed his belief these might be caused by the pressure of sunlight. He guessed right, and so did Halley in his letter to Newton. But a real understand-

41 An old woodcut showing Tycho's comet of 1577 with an astrological prediction based on its advent.

ing of the physical processes taking place inside comets and their tails awaited a much later date and astronomers of a future age. Halley, like Newton in the *Principia*, concentrated his researches on computing orbits of comets to see what these might explain...

Taking a cue from Newton, Halley returned to first principles, accepting nothing on trust. Even Newton's own work came under rescrutiny. Sifting his way through historical records, he gleaned all available observations of previous comet apparitions he could find. Most of these were worthless in terms of their use to calculate precise orbits. A few looked more promising. From the data he assembled he selected twenty-four comet apparitions and then, involving himself in what he later described as "immense labour", he drew up a short selective catalogue.

From time to time he gave progress reports to Newton. In one significant letter he wrote: "... I fell to consider that comet of 1680-1 which you have described in your book [the *Principia*], and looking over your catalogue of observed places, I find in that of 25th January 1681, there is a mistake of 20 minutes in the [celestial] Longitude of that day, or 56 minutes for 36, and so I have in a lre [letter] Mr Flamsteed sent me when I was in Paris. I though fitt to advertise this, because you wrote you designed to undertake to correct what you had formerly determined about the orb thereof; and that day is one of those you have taken to define the orb by..."

Newton must have been greatly annoyed to hear of this careless error contained in his masterwork which by now, among British cognoscenti, enjoyed an authority rivalling that of the Bible. It may have rankled him also to learn from Halley: "I find certain indications of an Elliptik orb in that Comet and am satisfied that it will be very difficult to hitt it exactly by a Parabolick. When I have computed all the observations, I shall send you what I have done."

Halley believed he was on the verge of a great discovery. His basic research tool remained the *Principia*. Newton had provided the apparatus but ironically the example used to illustrate its practical working was defective. Halley, as the editor and publisher of the *Principia* and the leading disciple of Newton, understood the author and his work better than any other man. In addition he had a flair for historical scholarship and ancient learning which was superior to that of Newton's. By diligent research he stretched his comet investigations back to before the time of Kepler and Tycho Brahe.

One of Halley's problems, just as Newton's had been when evolving his comet theory for the *Principia*, was obtaining observations of past apparitions reliable enough to use for calculating orbits. Without accurate observations, it was near impossible to decide the curve in which a comet moved. Halley knew that Flamsteed had made some excellent observations of the 1682 comet, but his problem now was how he was going to prize them loose from the man who discredited him at every opportunity.

Halley's current standing with Flamsteed was at a particularly low ebb. On 7 February 1695, Flamsteed had written to Newton: "I have done with him [Halley] who has almost ruined himselfe by his indiscreet behaviour & you shall hear no more of him from me till wee meet when I shall tell you his history which is too foule and large for a letter."

Whatever these "too foule and large" deeds were supposed to be, they do not appear to have influenced Newton's opinion of

42 The Octagon Room at Greenwich Observatory from where Flamsteed and his assistants observed the comet of 1682 (later Halley's comet).

43 Newton's manuscript notes for his observations of the comet of 1682 (Halley's comet). The accompanying sketch shows the shift of the comet on four separate nights against the backcloth of the Great Bear.

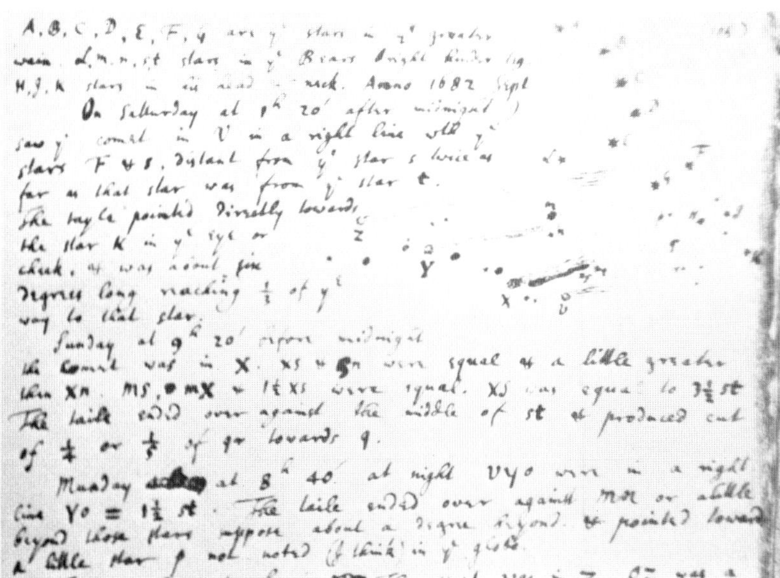

A, B, C, D, E, F, G are ye stars in ye greater
wain. L, m, n, s, t stars in ye Bears hinder leg
H, J, K stars in its head & neck. Anno 1682 Sept
On Saturday at 9h 20' after midnight I
saw ye comet in V in a right line with ye
stars F & s, distant from ye star s twice as
far as that star was from ye star t.
The tayle pointed directly towards
the star K in ye eye or
cheek. It was about five
degrees long reaching 1/3 of ye
way to that star.
Sunday at 9h 20' before midnight
the Comet was in X. XS & Xn were equal or a little greater
than Xn. MS, mX & XS were equal. XS was equal to 3½ st
The taile ended over against the middle of st & produced cut
of ¼ or ⅕ of qr towards q.
Munday at 8h 40' at night VYO were in a right
line YO = 1½ st. The taile ended over against mn or a little
beyond those stars opposite about a degree beyond. & pointed toward
a little star not noted (I think) in ye globe.
Tuesday at 9h 0 The comet was in Z. OZ was a

Halley. We may suppose they refer to idle coffee-house gossip respecting Halley's past affairs of the heart.

Truth is, Newton was rapidly losing patience with Flamsteed. He thought very little of the Astronomer Royal, and his contacts were based on self-interest, in particular securing Flamsteed's observations of the Moon so he might perfect his lunar theory for a new edition of the *Principia*.

Halley took the obvious, if devious, way out and wrote to Newton requesting he obtain Flamsteed's comet observations: "I must entreat you to procure for me of Mr Flamsteed what he has observed of the Comett of 1682 particularly in the month of September, for I am more and more confirmed that we have seen that Comett now three times, since ye Year 1531, he will not deny it you, though I know he will me."

Newton obtained the necessary observations, and Halley, on receipt of them, recognized their worth, offering generous praise to the man who had made them. He replied to Newton he wished Flamsteed had also made similar high-quality observations for the 1680-1 comet as all the other ones he had collected were from "very course observers".

Halley's intentions in pursuing his new comet-studies were to extend the scope of Newton's pioneer work—taking into account the consequences of the gravitational effect large planets may have on comets. Seeking guidance, he wrote to Newton: "I must entreat you to consider how far a Comet's motion may be disturbed by the Centres' of Saturn & Jupiter particular in its ascent [its retreating path from the Sun] and what difference they cause in the time of the Revolution of a Comett in its so very Elliptick Orb."

In requesting this information, Halley had again put his finger on the key to the problem of cometary orbits. As a consequence of Newton's universal gravitation, comets had been shown to be fully paid-up members of the cosmic club and no different from the planets in obeying the attractive dictates of other nearby bodies. The amount of pull was dependent on the mass (bulk) and proximity of the various bodies to each other. Halley recognized that in the case of comets moving in orbits which allowed them to pass close to massive bodies like Jupiter and Saturn, they might be subject to much greater influences than in close encounters with the smaller, terrestrial planets like Mars, Earth and Venus. The problem still unresolved was the nature of the comets themselves and their mass in relation to the planets.

Halley included in his study the Plague and the Fire comets he had seen over London as a boy. The path of the Plague comet presented him with his greatest problem. Hooke had claimed it was a new apparition of the 1618 comet but offered no proof it was. Halley was inclined to dismiss this idea. To calculate its orbit he tried using observations of it made by his old acquaintance Hevelius now dead for some eight years. He soon suspected that the wily old Danzig astronomer had cooked these observations, noting he had added "8 or 9 minutes to make his figures agree" to the places observed.

Halley, as well as Newton, was aware that in the very first number of the *Philosophical Transactions* was a paper by the French savant Adrien Auzout in which he attempted to predict the movement of the Plague comet in 1664. Auzout held a very high opinion of his own abilities and in a self-puff referred to his project as "a design which never yet was undertaken by any Astronomer, all the World having been hitherto persuaded that the motions of Comets were so irregular that they could not be reduced to any Laws".

Halley's goal was not simply to rationalize past observations and obtain orbits. He wished to make the great step forward and accurately predict, as astronomers already could do for the planets, the exact place in the heavens where a particular comet would be seen in the future. To do this he realized he must know whether a particular comet would be influenced unduly in its timetable by the proximity of Jupiter or Saturn. In such an instance, would a comet be held back in its course or, alternatively, accelerated?

The two comets which interested him most were the 1680-1 comet he had seen in all its magnificence on the road from Calais to Paris as it re-emerged from the direction of the Sun and the 1682 comet seen from Islington shortly after his marriage. In the case of the 1680-1 comet he discarded Newton's previous work and recomputed it, finding an elliptical orbit which indicated a period of revolution round the Sun of 575 years. Such an orbit would carry the comet into space far beyond the outermost planet Saturn, then the farthest known. Thus Halley concluded the solar system was a much vaster entity than astronomers had previously supposed it to be.

The results Halley obtained for the comet of 1682 were even more interesting. Its orbit again proved to be elliptical but this time not so elliptical as the 1680-1 comet. If his calculations were correct, the 1682 comet had a period of revolution round the Sun of 76 years...

In arriving at this figure, Halley had been guided by the knowledge that Kepler had observed a bright comet in 1607 and, previous to that, Peter Apian had seen one in 1531. There had also been a comet seen in 1456. The fingerprints, the elements—the nodes and perihelia—of their respective orbits were very similar. He discounted coincidence but to be absolutely sure he checked and rechecked his calculations.

Although Halley discussed these interesting findings at Royal Society meetings at this time, he wrote up no definitive paper about them. This was probably because outside events interfered to change the direction of his life.

In the early 1690s, Halley's interests had centred round sea-diving and a study of the Earth's magnetism. During the 1680s-90s there had been great public curiosity in sea voyaging. Since his own voyage to St Helena, Halley had retained a strong interest in the sea and only waited opportunity to go sailing again. Then in 1693, in conjunction with a friend, Benjamin Middleton—about whom practically nothing is known—Halley planned a scientific voyage. We know only the bare outline facts of the proposed enterprise from Admiralty and Royal Society documents and from Hooke's diary where he recorded that Halley "made proposals of going into ye South Seas & round the World".

Apparently the pair of would-be adventurers had drawn up a plan and then approached the Royal Society to lend assistance for "a small vessell of 60 Tuns" to be fitted out by the government, but to be victualled and manned at Halley's and Middleton's own charges "so as to encompass the Globe to make observations on the Magneticall Needle, Ec".

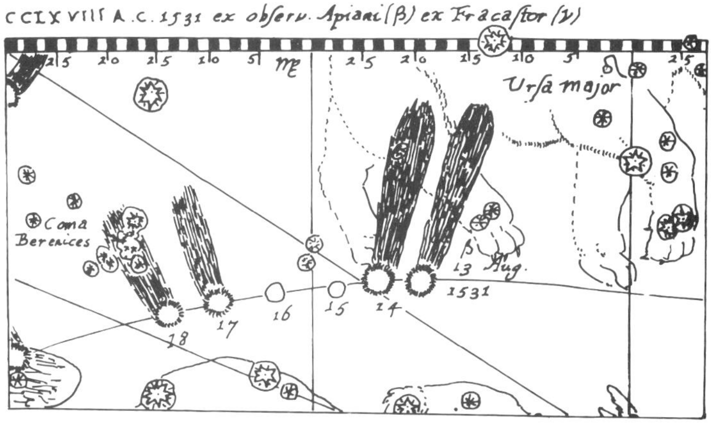

44 The path of Halley's comet in 1531 as plotted by the observations of Peter Apian in Ingolstadt and Fracastor in Verona (from *Theatrum Cometicum*).

45 The dramatic appearance of Halley's comet in 1456 when it spread itself across three constellations. The two bright stars near the head of the comet are Castor and Pollux in the constellation of the Twins. It was during this apparition the Turks were besieging Belgrade and the Pope offered prayers against the enemies of Christendom and the comet. (based on a depiction by an unknown artist)

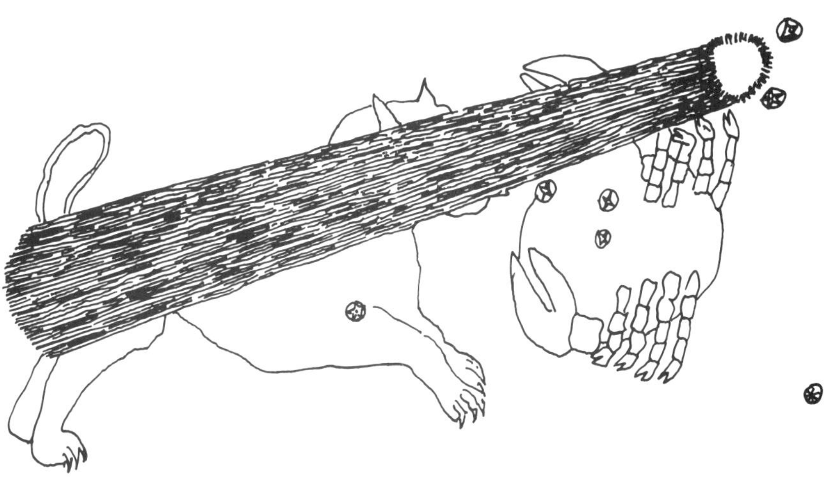

The Navy Board were favourably disposed to the proposals put before them. They had already been impressed with Halley's diving-work and his apparent familiarity with practical naval problems. By 1694 plans had taken shape. A ship, a Pink called *Paramore*, was to be constructed and made available to him, and it seemed as if Halley was all set for a new career as scientific voyager. The Clerk to the Society now received a royal commission appointing him master and commander of the *Paramore*, together with an authorization to equip the vessel with a regular complement of naval officers and crew.

Two years slipped by.

Why Halley and Middleton delayed their voyaging is not known with certainty, but it was probably due to delays in rigging the *Paramore* in combination with the protracted arrangements for victualling and payment of a crew. By 1696, when Halley was still busy with his comet researches, another unexpected interruption caused him to postpone his South Seas expedition indefinitely.

William III was firmly in power. The Parliament of the day had become concerned over the widespread practice of coin clipping whereby the entire economy of the country was threatened. Only coins impressed with a distinct milled edge were safe. It was decided to recoin, and Newton, among others, was asked to advise on how this should be done.

In 1693-4, Newton had undergone a serious depression and was constantly on the verge of a nervous breakdown. This is a shadowy period in his life, but his new involvement with the recoinage and his appointment as Warden of the Royal Mint in 1696, and then his appointment as Master in 1698, changed the entire course of his life. He entered into it with a new zest. In 1696, after residing at Trinity College for thirty-five years, he suddenly departed for London. His scientific career was over, but he did not resign his Lucasian professorship for another five years. During that time he returned to Cambridge just once and then it was only for a short visit lasting half a week.

Newton's first task was to organize the country mints—in Norwich, York, Bristol, Exeter and Chester. Soon after taking over the reins, he realized a pressing need for men of integrity to administer these establishments. The fact he had been in recent contact with Halley discussing comets probably influenced his decision to call upon his services. Perhaps he was aware he had never repaid his debt to Halley for his work on the *Principia*. Now, as a man of power and influence, he had opportunity.

By late 1696, Halley was ensconced in the Chester Mint as its deputy comptroller, drawing a salary of £90 per annum. Unlike Newton, however, who now acted like a man reborn and amazed all his friends with his unexpected flair for administration, Halley was unhappy in his new role. He carried out his duties diligently, but there were rivalries and squabbles, and soon he detected malpractices. At best, the work, described in his own words "was drudgery", but it was not the intellectual drudgery he was so well fitted for. Neither had he anticipated that those he accused of dishonesty would try to turn the tables on him. In the end Newton was called upon to protect him. To his credit, Newton—roaring like a lion—defended his old friend before the Lords' Commissioners and carried the day. Grateful for Newton's patronage and help in clearing his name, Halley agreed to remain at Chester until the country mints were closed down in 1698.

The episode in the Chester Mint showed Halley would never be fitted for a life in commerce. Newton, by contrast, had performed brilliantly, and the academic life of a don at Cambridge was all to willingly exchanged by him for that of an influential man-of-the-establishment in London. A relentless pursuit of the forgers and their execution to the last letter of the harsh English laws now seemed a good deal more satisfying to his intellect than contemplating the laws of nature. In 1705, as a worthy and willing tool of the administration, he was knighted by Queen Anne for services rendered.

As for Halley—after Chester—voyaging was now in the air again. Under the date of 15 October 1698 the Admiralty, on Halley's own recommendation, issued instructions for a twelve-month scientific voyage designed to investigate and improve knowledge of magnetic variation.

CHAPTER 10

Scientific Voyager

"You are to make the best of your way to the Southward of the Equator and there observe on the East Coast of South America, and the West Coast of Africa, the variations of the Compasse, with all the accuracy you can... you are likewise to make like observations at as many of the Islands in the Seas between the aforesaid Coasts as you can... and if the season of the Year permit, you are to stand soe farr into the South, Till you discover the Coast of Terra Incognita..."

—Excerpt from Admiralty Instructions to Halley

Halley must have been relieved when the Admiralty confirmed they were still interested in his proposals. His Chester experience did little to further his public reputation as an administrator and manager of men. If the Sea Lords had doubts about their appointee's recent experiences at the Mint, they waived them, for they must have recognized Halley's undoubted abilities as a scientist and as a man of integrity.

Halley must have keenly anticipated the prospects of pursuing the search for Terra Incognita, the legendary great land which lay somewhere to the far south. Its sure delineation was a geographical prize sought by all voyagers to the South Seas. The Dutch East Indiamen on their voyages round the Cape to the Spice Islands had sometimes made tantalizing landfalls along the coasts of Western Australia, but no one was yet sure how far this great continent extended; some maps of the period showed Terres Australes sprawled round the entire southern polar regions.

Before Halley set out on what was to be his first voyage, there were further delays and more diversions. By this time Middleton's name is never mentioned as a co-voyager, and the reason for his dropping out is not known (*see* note 8). A diversion earlier in the year was caused by Czar Peter, who arrived in

London as the guest of William and Mary to study shipbuilding at first hand. According to some stories he asked to be put in touch with England's most knowledgeable men. If de Mairan's éloge of Halley's life and a biographical note by Martin Folkes are correct, the English astronomer was apparently one of the chosen few. However, there is little evidence to throw light on the nature of their relationship, but Martin Folkes tells us: "This Great Prince was highly pleas'd with him [Halley], and treated him with great distinction, admitting him to the Familiarity of his Table, to have the more opportunitys of being inform'd by his enter [entertaining] and instructive Conversation."

There is no reason to doubt Halley did actually meet with Peter, but in all *contemporary* accounts Halley's name is not once mentioned in juxtaposition with the Czar's. Their meeting likely came about because of Peter's interest in the *Paramore*, the vessel to be used for Halley's long-postponed voyage. We know Peter made a request that she be rigged and made seaworthy so he could carry out sailing experiments with her and no doubt Halley was kept informed about "his frigatt", as he refers to her while he was still at Chester, and came down to London to supervise the operation. Whether as a result of their meeting Halley became involved in Peter's wild drinking-sprees is conjectural. Yet in view of Flamsteed's later remarks, a few years hence, about Halley's alleged new fondness for brandy—which supposedly was introduced into London by the Czar—we might readily speculate it was Peter who acquainted him with it.

The sight of London that doubtless would have fascinated Halley and likewise Peter was the Thames anchorage of merchant ships—known as the Pool of London. It was about this time, or possible a little later, that Daniel Defoe once counted upwards of 2,000 ships moored there in rows.

It was from the Pool that Halley set sail on his first voyage about noon on Thursday 20 October 1698. By now the sometime gentleman astronomer, Clerk to the Royal Society, then Deputy Comptroller at the Chester Mint was officially transformed into "Captn Edmd Halley Commandr. of his Mats. Pink Paramour". He seems to have taken the transition to sea captain in his stride. Apart from his voyage (as a passenger) to St Helena, over twenty years before, and then his latterday off-shore experiments with diving-bells and ships' rigging, he had little first-hand nautical experience. Certainly the professional naval captains, the "tarpaulins", as Pepys used to call them, would resent the opportunity given to the gentleman amateur totally lacking in practical sea experience.

It is hardly surprising this same resentment soon infected the regular naval officers seconded to the *Paramore* (or *Paramour* or *Parrimore* as she was sometimes called in documents), and before long it infected the lower-deck crew as well.

Discontent with the amateur sailor festered slowly until it finally manifested itself by a perverseness to frustrate the captain's orders when he was attempting to make a landfall at Fernando de Noronha off the west coast of Brazil. But Halley was not to be thwarted; he was not going to turn a blind eye and let disobedience pass comment until it was too late to correct it which had been partly the cause of his troubles at the Chester Mint. He immediately exercised his authority, and next morning they made the required landfall and secured his anchorage. This, however, according to his account, was not without some danger to the *Paramore*, "having narrowly escaped a Sunk Rock".

After further adventures along the coast of Brazil, including trips ashore to make astronomical and magnetic observations, they reached Barbados in April. It was now resentments among the officers simmered over again. The first lieutenant, a man called Edward Harrison, who had an old score to settle with Halley, was the chief culprit. Eventually he became so impertinent that Halley placed him under arrest. The voyage was abandoned, and Halley assumed sole charge of navigating the *Paramore* back to Plymouth where they arrived on 23 June 1699. It was left to their Lordships at the Admiralty to decide Harrison's fate.

Halley readily admitted his lack of experience: "perhaps I have not the whole Sea Dictionary so perfect as he ... he for a long time made it his business to represent me, to the whole Shipps company, as a person wholly unqualified for the command their Lordships have given me, and declaring that he was sent on board here because their Lordships knew my insufficiency."

Harrison had sorely affronted Halley's pride; he had denounced him openly before the other officers, claiming the Captain did not have the capacity of commanding a ship's long-boat let alone a vessel like the *Paramore*. Stung by this unfair impeachment, Halley was determined to teach his mocking first lieutenant a lesson in manners and loyalty. Harrison could not have expected Halley to place him under arrest as he would have reckoned himself indispensable for the safe navigation of the ship. Overcome by arrogance, he was blind to the fact that Halley was much his superior when it came to the art of finding a ship's way by the Sun, Moon and stars.

Halley had proved his point. All along he had been aware of the root cause of Harrison's resentment and his truculence in accepting orders from him. Back in 1696, Harrison had published a small volume, *Idea longitudinis*, in which he proposed a method of determining accurate longitude at sea. Harrison's book was submitted to a pool of experts for their assessment. One of the experts was Halley. Their consensus findings were that they thought little of Harrison's ideas—believing them quite impractical. Harrison had been unable to forget this slight to which he believed Halley had been a major party. Whether Harrison had purposely schemed his move for secondment to the *Paramore* is uncertain. If not, it seems a remarkable coincidence that he found himself acting as Halley's second-in-command in a situation where he might find himself well placed to show the amateur captain in poor light.

Harrison in due course faced a court-martial but survived it with only a stern reprimand—a light sentence indeed for a crime bordering mutiny. He was able to prove that, technically, he had not actually disobeyed Halley's order, and the only charge to stick was one of insolence. One can imagine that all the regular naval officers comprising the court-martial would have been in sympathy with Harrison's case and therefore decided to let him off lightly, thus providing their own snub to the amateur captain. The Admiralty, however, fortunately recognized the incident in its true light and retained their faith in Halley, who was already making preparations for a second voyage.

Before his first voyage in 1698, Halley likely encountered the ex-buccaneer William Dampier, who was preparing for a new voyage to explore the coasts of Australia and New Guinea at the instigation of the Admiralty. Round this time the colourful Dampier dined several times with Pepys, and it was probably at one of these dinner parties the two mariners first met face to face. In spite of his past reputation, Dampier had a lot in common with Halley. He was a self-taught naturalist and observer of nature with strong interests in navigation and magnetism. On his new voyage he undertook to make magnetical observations in the regions not covered by Halley. During the two years' duration of Dampier's voyage in *Roebuck* he collated a table of observed variations of the compass at a large number of locations later published in his *A Voyage to New Holland*. Dampier paid generous tribute to Halley in his book, remarking: "I cannot but hope that the ingenious Author, Capt. Halley, who to his profound skill in all theories of these kinds hath added and is adding continually Personal Experiments, will e'er long oblige the World with a fuller Discovery of the Course of the Variations, which hath hitherto been a Secret." (*see* note 9)

Halley departed England on his second voyage on 27 September 1699. It got off to an inauspicious start when the pumps of the *Paramore* became clogged with sand. Then in October, when the vessel was near the Canaries, one of Halley's favourites, a deck boy called Manley White, fell overboard and was lost. It was an event which cut Halley deeply. Friends related that years later, whenever he recalled the incident, a tear would fall from his cheek. Young Manley was the only loss of human life suffered during Halley's three voyages as a commander.

This time the aims of the scientific voyage were not to be thwarted. By 27 January the following year *Paramore* had reached south latitude 51°, near the Antarctic Convergence. Halley's log is peppered with descriptions of the prolific sea-life surrounding the vessel: vast herds of seal, schools of penguin and flights of shags constantly presented themselves. They also sighted killer whales, not quite knowing what kind of creatures they were. Halley's log relates: "some supposed [them] to be Seals but are not soe; they bent their Tayles into a sort of Bow ... and being disturbed shew'd very large Finns as those of a large

Shirk [shark]. The head not much unlike a Turtles."

Now in high southern latitudes, among unknown seas with the ice-bound Terra Incognita lying just beyond the horizon, the temperatures dropped to a little above freezing. Halley tells us he allowed his crew their full ration of rum to help them keep warm when working the ship.

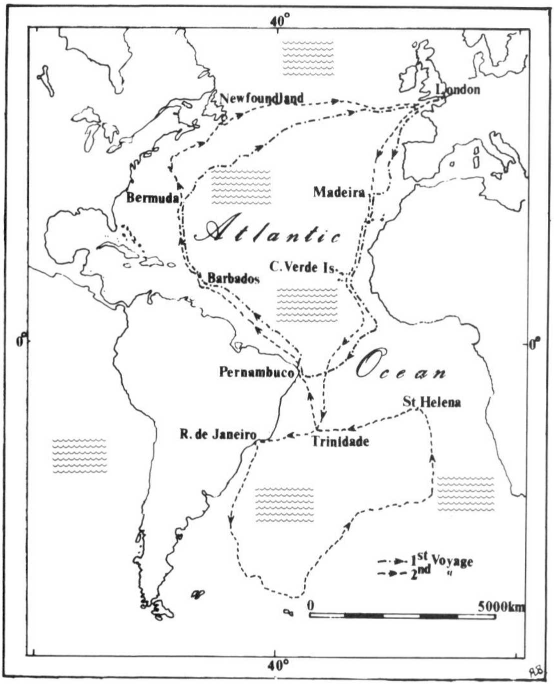

46 The track of the *Paramore* during Halley's first two voyages 1698–1700.

As they pushed farther south, the temperature fell below freezing point, a sure indication of potential danger to the mariners. Now they sighted their first great tabular icebergs and, perhaps not surprisingly, in their ignorance mistook them for new islands "... being all flatt on Top, and covered with snow. milk white with perpendicular cliffs all round... The great hight of them made us conclude them land, but there was no appearance of any tree or green thing on them, but the Cliffs as well as topps were very white".

Oblivious to possible danger and curious to discover the

nature of the new "undiscovered" islands, Halley pushed the *Paramore* ever closer. He was now courting disaster, but in his ignorance was unware of it. *Paramore* was no vessel in which a captain could take liberties for she displaced only 89 tons and was only 52 feet long by 18 feet wide with a draft of 7 feet 7 inches. Suddenly, to his horror, he realized what these new islands really were. He was astounded by the revelation that they were "very distinctly to be nothing else but one body of Ice of an incredible hight whereupon we went about Shipp and stood to the Northward".

The following day Halley added more to his experience: this time about the hazards of ice combined with fog, a common occurrence round the Antarctic Convergence. But it was an experience gained at some peril to the *Paramore*. "Between 11 and 12 this day," he relates, "we were in iminant danger of loosing our Shipp among the Ice, for the fogg was all the morning so thick, that we could not see for long about us, where on a Sudden a Mountain of Ice began to appear out of the Fogg ... another appeared more on head ... and had not beene halfe a quarter of an hour under way when another mountain of Ice begins to appear on our Lee bow ... but the sea being smooth and the Gale fresh wee got Clear: God be praised. This danger made my men reflect on the the hazards wee run. in being alone without a Consort, and of the inevitable loss of us all in case we staved our Shipp which might soe easily happen amongst these mountains of Ice in the Fogg, which are so thick and frequent there."

James Cook, who visited these same waters three-quarters of a century later, encountered similar near-disasters during his own search for the still elusive Terra Incognita.

Repulsed by the ice mountains and fog, Halley sailed *Paramore* north-eastwards so as, according to his log, to "recover the Warm Sunn, who we have scarce seen this fortnight". Nevertheless, the cold, foggy damp penetrated everywhere, and he complained bitterly: "our linen, our Cloaths, our papers & c feel wett with it, even in our Cabbins."

Free of the bergs, they encountered a "Terrible high Sea" which at one stage threated to broach the *Paramore* "but please God, she wrighted again".

Halley's search for Terra Incognita was over; after paying a call to his old haunt at St Helena, he sailed northwards, mapping variations of the compass, the chief purpose of the voyage, as he progressed. There was a brief adventure at Recife where the British consul was suspicious that Halley was a pirate! The consul was not satisfied until Halley provided incontrovertible proof that *Paramore* sailed under Admiralty orders.

Halley's original orders had included a call at the West Indies on his return home. He was instructed to visit as many of the islands as possible and "to make such observation as may contribute to lay them downe truly in their Geographical Scituation". At Barbados a more serious incident occurred when Halley and his crew were suddenly struck down by "a severe pestilentiall dissease ... my selfe and many of my men were seazed with it ... which scarce spares any one".

As soon as he and his crew recovered, they continued sailing northwards, running along the North American shore past Nantucket and then Newfoundland where they suddenly surprised some English fishing-boats off the Grand Banks. There must have been a strange, rather sinister look to the *Paramore's* lines, for Halley tells us: "they all fled from us takeing us, as we found afterwards, for a Pyrate."

Halley navigated his ship back across the North Atlantic and in August 1700 reached England without further incident. His second voyage to the South Atlantic had proved a great success. Although he was unable to reach the coasts of Terra Incognita, the scientific work accomplished was outstanding. From it he later published a large Atlantic chart which pioneered the method of depicting magnetic variations. The Admiralty were well satisfied with their investment, and his captaincy had been vindicated. Now, before he retired from the sea to resume his academic studies, their Lordships had a very special commission for him. On 14 June 1701, Halley sailed from Deptford again aboard *Paramore* with official instructions to determine the tides and chart the headlands and promontories of the English Channel.

There seems doubt that on her third voyage *Paramore* also sailed under more secret orders as an incognito spy ship. Martin

Folkes, a reliable source of contemporary and very gossipy Halleiana, believed the new Channel survey was a cover for more clandestine activities. At this time British relations with France were at a low ebb, and apparently very little intelligence was known about the strength of the French defences. Halley's task then, under the guise of a third scientific voyage, was to obtain military information about the French dispositions round their Channel ports.

By 10 October 1701, Halley had finished his Channel "survey" and was back at Deptford. After handing *Paramore* over to the Port Captain, his brief but adventurous naval career was at an end.

Not long after, the War of the Spanish Succession soon had Halley in government service again and under a guise not so different from that of his Admiralty brief for the third voyage.

Affairs in Europe in the early 1700s were tense. In September 1701, Leopold of Austria entered into the Grand Alliance with the Engish and their old adversaries the Dutch. In the name of Queen Anne, who succeeded to the throne in March 1702, a very special emissary was required to travel to Europe, and Halley's name was put forward as the man with the experience and qualities required. State papers shed little light on the circumstances, but it was very likely Dr John Wallis, Savilian Professor of Geometry at Oxford—a cipher expert and a man deeply involved at the top in secret-service activities—who recruited Halley. Armed with a special letter of authority, Halley sailed from London to Holland on 27 November 1702 and then from there made his way to Vienna. Here his official brief was to advise the Emperor on the fortifications of harbours along the northern shores of the Adriatic.

There was a strong tradition during the Renaissance and the sixteenth and seventeenth centuries for men of science to be experts in warfare and weaponry as well. Galileo was a prime example of this. Others had refused such commissions, believing science and warfare were uneasy bedmates. For this reason Christopher Wren turned down Charles II's offer to supervise new defences in Tangiers. Newton also was openly hostile to the application of science to warfare, but not so Halley whose keen interest in nautical and naval matters had given him opportunities to further his career.

47 Halley c 1700 dressed in the uniform of a naval captain. The original by Sir Godfrey Kneller is now lost and the portrait exists only as an engraving by G White (16??–1732).

In Vienna the Emperor and his advisers are said to have been well pleased with Halley's expertise, and he was presented with a diamond ring of great value for services rendered. After returning to England briefly, he was soon off again, this time to supervise the building of all the necessary works he had recommended in his report.

Halley was now moving in high circles like his compatriot Newton. During his journey to Vienna he stopped over in Hanover where he was entertained by the Queen of Prussia and the Electoral Prince, later to become George I of England. In Vienna he was again entertained by the Emperor and treated as a VIP.

Documentary evidence which sheds light on the details of Halley's mission is tantalizingly scarce. However, one indicator about his clandestine activities is that reimbursements later made to the incumbent British minister in Vienna for local expense monies he had paid out to "Captain Halley" were funded "out of secret service". Unfortunately state records so far examined shed little further light on Halley's sometime 007 activities (*see* note 10).

CHAPTER 11

Synopsis Astronomiae Cometicae

In October 1703, a month before Halley returned to London from Vienna, Dr John Wallis died, and a vacancy was declared at Oxford. Halley's friends immediately put forward his name. His old adversary Bishop Stillingfleet was no longer a threat as he had died four years previously. Neither was Bentley, the bishop's ex-chaplain and fellow conspirator to keep Halley from academic office in 1691, a force to be reckoned with. As Master of Trinity, Bentley had his own troubles and was immersed in an unbroken series of quarrels and litigations owing to his "arrogance and rapacity". Nevertheless, the old enemy Flamsteed was still prepared to intercede and attempt to spoke Halley's wheel. In December 1703 he wrote to his ex-assistant Abraham Sharpe: "Dr Wallis is dead. Mr Halley expects his place. He now talks, swears and drinks brandy like a sea captain, so that I much fear his own ill-behaviour will deprive him of the vacancy."

Halley, however, was riding on the crest of his newly won reputation as a man of some import. He had influence both at court and in the government. For more than one reason he was considered a worthy successor to the man who had recruited him for service abroad. Nothing that Flamsteed might do could deprive him of the vacant professorship for which he was by far the best candidate. He and Newton were on excellent terms; Newton was President-elect of the Royal Society and one of the most influential men in England. This time he gave Halley his full public support when his name was put forward as Wallis's successor.

On 30 November 1703, Halley was elected to the Council of the Royal Society, gaining more votes in the ballot for officers than Newton had done. Then, with very little opposition, he was duly elected to fill the Savilian Chair of Geometry at Oxford. It was a proud moment for him when in May 1704 he returned to his Alma Mater to deliver his inaugural lecture.

Halley was now aged forty-eight. He was in his prime, secure in the knowledge he had gained his own sheltered cloister where he could immerse himself in the intellectual work yet unfinished. Until his two honorary doctorates awarded a few years later he was more usually addressed as "Captain Halley", and one can imagine that as a scholar he was rather proud of this title and the reputation won as a man of action.

Halley was soon involved in the unfinished problem of the comets. Much of the drudgery of calculation had been done before he had gone to the Chester Mint in 1696. Ensconced in Oxford, he took opportunity to revise and check the earlier work and then prepared it for the University Press. In 1705 he published what was to be his most important work under the title *Synopsis Astronomiae Cometicae (A Synopsis of the Astronomy of Comets).*

As the title suggests, it was a short book. In sharp contrast to Newton's 500-odd-page *Principia* it was more of a tract, only double the length of Newton's original *De motu*. Its eighteen small quarto pages were printed in Latin as was still custom, and on its cover was illustrated the Sheldonian Theatre, designed by his old friend and colleague Wren. An edition written in English was published simultaneously in London and it was also reproduced in *Philosophical Transactions.*

In his *Synopsis* Halley accomplished the historical introduction in three succinct pages. The kernel of the whole work was contained in a table, presenting the (parabolic) elements of the twenty-four comets he had singled out for investigation (Table 1). To calculate these orbits in a pre-computer age had involved him in the most gruelling tedium. Newton, almost twenty years before, had provided the method and insights. Now, as the leading disciple and by heroic application, Halley was able to reap the subsidiary harvest that had eluded the author of the *Principia*.

48 Halley's *Synopsis of the Astronomy of Comets* (1705) was printed in both English and Latin versions and subsequently appeared in supplements to works by others.

A SYNOPSIS OF THE Astronomy of COMETS.

By EDMUND HALLEY L.L.D. *Savilian* Professor of Geometry at *Oxford*.

THE ancient Egyptians and Caldeans (if we may credit *Diodorus Siculus*) by a long course of observations, w[illegible] said to be able to predict the Apparition[illegible] mets. But since they are also said [illegible] of the same Arts, to have pr[illegible] quakes and Tempests, [illegible] their knowledge [illegible] rather of m[illegible] any A[illegible]

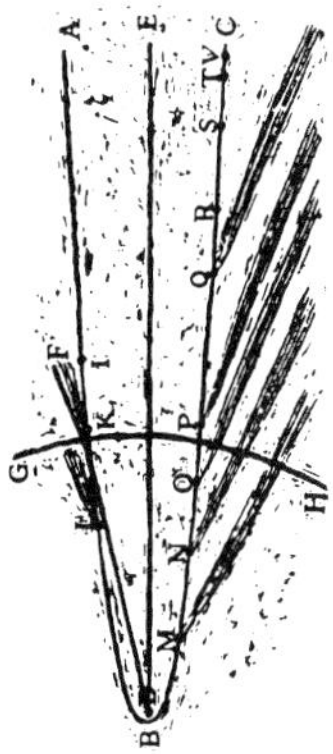

was bent into a Curve Line concave towards the Sun. At length, came that prodigious Comet of the Year 1680. which descending (as it were from an infinite Distance) perpendicularly towards the Sun, arose from him again with as great a Velocity.

This Comet, (which was seen for four months continually) by the very remarkable and peculiar Curvity of its Orb (above all others) gave the fittest occasion for investigating the Theory of its Motion. And the Royal Observatories at *Paris* and *Greenwich* having been for some time founded, and committed to the care of most excellent Astronomers, the apparent Motion of this Comet was most accurately (perhaps as far as humane skill cou'd go) observ'd by Mrs. *Cassini* and *Flamsteed*.

Not long after, that great Geometrician the Illustrious *Newton*, writing his *Mathematical Principles of Natural Philosophy*, demonstrated not only that what *Kepler* had found, did necessarily obtain in the Planetary System; but also, that all the Phænomena of Comets wou'd naturally follow from the same Principles; which he abundantly illustrated by the example of the aforesaid Comet of the Year 1680. shewing at the same time, a method of delineating the Orbits of Comets Geometrically; therein solving (not without meriting the highest admiration of all Men) a Problem, whose intricacy render'd it scarce accessible to any but himself. This Comet he prov'd to move round the Sun in a Parabolical Orb, and to describe Area's (taken at the Center of the Sun) proportional to the Times.

Wherefore (following the steps of so Great a Man) I have attempted to bring the same Method to Arithmetical Calculation; and that with all the success I cou'd wish. For, having collected all the observations of Comets I cou'd, I have fram'd this following Table, the result of [illegible]rodigious deal of Calculation: which, tho' [illegible]ll in bulk, will be no unacceptable pre[illegible] [illegible]onomers. For these Numbers are [illegible]esenting all that has been yet [illegible] Motion of Comets, by the [illegible] General Table; in the [illegible] labour, that it might [illegible] consecrated to [illegible]stronomy it

(1807)

tis, ut distantia Aphelia in Ellipsi est ad Axem totum Ellipsis; Velocitates autem sunt in dimidiata ratione eorundem: quapropter in Orbibus valde Excentricis ratio hæc accedit proxime ad rationem æqualitatis. Tantilla autem differentia, quæ intercedit ratione majoris in Parabola velocitatis, facillime in situ Orbis determinando compensatur. Hujus itaque Tabulæ Elementorum Motuum usus præcipuus est, atque etiam propter quem illam construere operæ pretium duxi, ut, si quando novus Cometa emerserit, possimus collatis elementis dignoscere an poterit esse aliquis ex antiquis, necne; ac proinde Periodum Orbitæque Axem determinare, reditumque prædicere. Ac sane multa me suadent ut credam Cometam anni 1531 ab *Apiano* observatum, eundem fuisse cum illo qui anno 1607 descriptus est à *Keplero* & *Longomontano*, quemque ipse iterum reversum vidi ac observavi anno 1682. Quadrant Elementa omnia, ac sola inæqualitas periodorum adversari videtur: hæc autem tanta non est ut causis Physicis non possit attribui. *Saturni* enim motus à cæteris, præsertim *Jove*, ita interturbatur, ut per aliquot dies integros incertum sit hujus Planetæ tempus Periodicum. Quanto magis talibus erroribus obnoxius erit Cometa, qui quatuor pene vicibus altius excurrit *Saturno*, cujusque velocitas, vel tantillum aucta, Orbem ab Elliptico in Parabolicum possit immutare? Confirmatur etiam eundem esse potuisse ex eo, quod anni 1456 æstate, conspectus fuerit Cometa eodem pene modo inter Solem & Terram transiens retrogradè: quem, licet à nemine observatus fuerit Astronomicè, ex periodo modoque transitus non diversum a prædictis extitisse conjicio. Unde ausim ejusdem reditum fidenter prædicere, anno scil. 1758. Quod si hoc evenerit, nulla amplius erit dubitandi causa, quin redire debeant cæteri. Habebunt ergo Astronomi in hac arenâ quo se exerceant per multa Secula, priusquam tot tantorumque Corporum circa commune centrum Solis revolventium numerus cognoscatur, ac motuum symptomata certis regulis coerceantur. Crediderim equidem Cometam

In his *Synopsis* Halley wrote: "... Hitherto I have considered the Orbits of Comets as exactly Parabolick; upon which Supposition it would follow that Comets being impell'd towards the Sun by a Centripetal Force, descend as from Spaces infinitely distant, and by their Falls acquire such a Velocity as they may again run off into the remotest Parts of the Universe, moving upwards with such a perpetual Tendency, as never to return again to the Sun. But since they appear frequently enough, & since none of them can be found to move with a Hyperbolick Motion, or a Motion swifter than what a Comet acquire by its Gravity to the Sun, 'tis highly probable that they rather move in very Excentric Orbits and make their Returns after long Periods of Time: For so their Number will be determinate, and, perhaps, not so very great.

"... And indeed, there are many Things which make me believe that the Comet which Apian observed in the year 1531 was the same with that which Kepler & *Longomontanus* took notice of and describ'd in the Year 1607, and which I my self have seen return, & observ'd in the Year 1682. All the Elements agree, and nothing seems to contradict this my opinion besides the Inequality of the Periodick Revolutions: which Inequality is not so great neither, as that it may not be Owing to Physical Causes. For the motion of saturn is so disturbed by the rest of the Planets, especially *Jupiter*, that the Periodick Time of that Planet is uncertain for some whole Days together. How much more therefore will a Comet be subject to such Errors, which rises almost Four times, higher than Saturn and whose velocity, tho' encreased but very little, would be sufficient to change its Orbit, from an Elliptical to a Parabolical one, ..."

The principal message then contained in Halley's little book was that the comet seen in 1682 was moving in a path only a little inclined to that of the major planets, and its orbit, in the shape of an extended ellipse, took it round the Sun in a period of about 75 years and this would bring it back into view sometime in the years 1758-9.

In 1705 the predicted reappearance of the comet was over half a century away. Halley was forty-nine years old and he would be dead when the comet reappeared; so too would most of his peers

Table I **Parabolic Elements of 24 Comets** (after Halley)

Date of Perihelion Passage	*Ascending Node*	*Inclination*	*Motion*	*Perihelion Longitude*	*Perihelion Distance*
1337 June 2 6 25	84° 21′ 0″	32° 11′ 0″	R	37° 59′ 0″	0.40666
1472 Feb. 28 22 23	281 46 20	5 20 0	R	45 33 30	0.54273
1531 Aug. 24 21 18	49 25 0	17 56 0	R	301 39 0	0.56700 *
1532 Oct. 19 22 12	80 27 0	32 36 0	D	111 7 0	0.50910
1556 April 21 20 3	175 42 0	32 6 30	D	278 50 0	0.46390
1577 Oct. 26 18 45	25 52 0	74 32 45	R	129 22 0	0.18342
1580 Nov. 28 15 0	18 57 20	64 40 0	D	109 5 50	0.59628
1585 Sept. 27 19 20	37 42 20	6 4 0	D	8 51 0	1.09358
1590 Jan. 29 3 45	165 30 40	29 40 40	R	216 54 30	0.57661
1596 July 31 19 55	312 12 30	55 12 0	R	228 16 0	0.51293
1607 Oct. 16 3 50	50 21 0	17 2 0	R	302 16 0	0.58680 *
1618 Oct. 29 12 23	76 1 0	37 34 0	D	2 14 0	0.37975
1652 Nov. 2 15 40	88 10 0	79 28 0	D	28 18 40	0.84750
1661 Jan. 16 23 41	82 30 30	32 35 50	D	115 58 40	0.44851
1664 Nov. 24 11 52	81 14 0	21 18 30	R	130 41 25	1.02575
1665 April 14 5 15	228 2 0	76 5 0	R	71 54 30	0.10649
1672 Feb. 20 8 37	297 30 30	83 22 10	D	46 59 30	0.69739
1677 April 26 0 37	236 49 10	79 3 15	R	137 37 5	0.28059
1680 Dec. 8 0 6	272 2 0	60 56 0	D	262 39 30	0.00612
1682 Sept. 4 7 39	51 16 30	17 56 0	R	302 52 45	0.58328 *
1683 July 3 2 50	173 23 0	83 11 0	R	85 29 30	0.56020
1684 May 29 10 16	268 15 0	65 48 40	D	238 52 0	0.96015
1686 Sept. 6 14 33	350 34 40	31 21 40	D	77 0 30	0.32500
1698 Oct. 8 16 57	267 44 15	11 46 0	R	270 51 15	0.69129

* Halley's comet

at the Royal Society. Halley's claim was a remarkable one but could not be proven or disproven until the passage of time permitted it.

Newton was very impressed with the thoroughness of Halley's work. In his revisions to the *Principia*, Newton was convinced enough to accept Halley's figure of 575 years for the period of the 1680-1 comet. In his second edition published in 1713, Newton included a large section devoted to cometary phenomena with special reference to the 1680-1 comet.

It is now known that Halley's figure of 575 years for the 1680-1 comet is wrong. Halley had used much the same observations as Newton when he had originally concluded the comet moved in an open-ended parabolic orbit. While Halley was undoubtedly correct in assuming the 1680-1 comet moved in an ellipse, it was an ellipse, nevertheless, with a much greater eccentricity than he could ever have imagined. In 1818 the German astronomer Johann Franz Encke made a detailed study of this comet and found its period of revolution was open to several possibilities. Because of the built-in inaccuracies of the observations made by Newton, Halley and their contemporaries, the comet's motion might be represented alternatively by an elliptical orbit of 8,800 years, one of 805 years, or possible by a hyperbolic one. The latter would imply that the comet after approaching the Sun and passing round it would then retreat into the depths of space never to visit the Sun again. However, the problem of true hyperbolic orbits for certain comets is still a contentious one even at the present time.

Encke also computed an alternative orbit for the 1680-1 comet. Using only Newton's and Flamsteed's probably more accurate observations, and rejecting Halley's and those made by others, he derived an elliptical orbit which gave a period of 3,164 years. Even so, it is clear that the true path of the 1680-1 comet is still a very uncertain one. Based on Halley's erroneous figure of 575 years, this comet, rather ironically, has been cited ad nauseam in many Earth catastrophe scenarios to the point when it has become a cliché (*see* note 11).

In the following years at Oxford, Halley settled down to a programme of solid academic work, writing numerous papers on all manner of topics. He also translated and reconstructed some of Apollonius's lost works previously known only from second- and third-hand Arab and Byzantine sources. In 1710 he was honoured by the University with the doctorates of LLD and DCL and from then on he was usually referred to as Dr rather than Captain Halley. His relationship with Flamsteed did not improve. By now Newton, as President of the Royal Society, was also at loggerheads with the Astronomer Royal over his tardiness in submitting observations to the Society from Greenwich as he was required to do. Following his breach with Newton, Flamsteed was blaming Halley, writing: "Mr Halley has set the Master of the Mint [Newton] at distance from me by false suggestions."

The final "insult" to Flamsteed was when Halley was given the

49 Portrait of Halley in early maturity.

task, as editor, of publishing the Greenwich star catalogue—the *Historia Coelestis*—which had languished in the Astronomer Royal's hands for years. After publication, Flamsteed—with some justification perhaps—tried to have it suppressed because of inaccuracies, calling it "a sacrifice to heavenly truth".

Past biographers of Flamsteed have sometimes quite unjustly maligned Halley for his supposed part in this particular affair. However, the relevant correspondence, not always accessible in the past, reveals Flamsteed to be the cause of all the bad blood as well as showing he was sometimes his own worst enemy. Surviving letters clearly show that Halley, in spite of Flamsteed's constantly hostile attitude towards him, tried to play fair. In a letter to Flamsteed dated June 1711, a year before publication of the *Historia Coelestis*, Halley writes in a friendly vein: "I have now sent the catalogue of the fixed stars intended to be prefixed to your book; having spared no pains to make it complete and correct as I could, by help of the observations you have given us, made before the year 1706. I desire you find all the real faults you can, not as believing there are none, but being willing to have a work of this kind as perfect as possible..." Then, anticipating Flamsteed's venom and perhaps hoping he might see reason, Halley adds: "and if you signify what's amiss, the errors shall be noted, or the sheet reprinted, if the case require it. Pray govern your passion, and when you have seen & considered what I have done for you, you may perhaps think I deserve at your hands a much better treatment than you for a long time have been pleased to bestow on your quondam friend, and not yet profligate enemy (as you call me)."

In the interests of his *Historia Coelestis*, which represented the cream of Flamsteed's life's work at Greenwich, here was an opportunity for him to accept an olive branch or at least a moratorium from his former protégé. But the Astronomer Royal would have none of it. Apparently he now wished to be one of science's martyrs. As a consequence, errors remained uncorrected—all due to Flamsteed's pigheadedness and refusal to co-operate.

In 1713, Halley was elected as one of the Secretaries to the Royal Society. Correspondence dated 1716 shows Flamsteed as tardy and hostile as ever in submitting observations to the Society as required by the rules set down by the Board of Visitors. Then, on 31 December 1719, Flamsteed died. He may have turned in his grave when Halley's name was put forward "as the fittest Person to do Honour to the Post of Astronomer Royal... which his Majesty George I was accordingly pleas'd to give him".

Halley remained in office as Astronomer Royal as well as retaining the Savilian Professorship at Oxford until he died in 1742. He was not a particularly memorable Astronomer Royal. By the time of his appointment his energies had diminshed, and this period is another very shadowy part of his life. Thomas

Hearne, who often met the astronomer day to day in Oxford, relates Halley had ambitions to relinquish his Professorship in Geometry and try for one in Geography, but nothing ever came of this. He also relates Halley had ambitions to complete a new edition of Ptolemy's classic *Geography* but that on hearing someone in Europe was bent on the same thing gave up the project after completing two volumes. He was active in Royal Society affairs to the very end. Apart from a stroke in the latter part of 1736, which caused him a slight paralysis in the right hand, he enjoyed excellent health. In advanced age, however, we hear that the old man "never eat a Thing but Fish, for he hath no Teeth".

Newton had died in 1727; Halley's 47-year-old friendship with him was a strange one. It is doubtful if they were ever on really intimate terms. Politically they were apart: Newton was a Whig while Halley, at heart, was a Tory. Newton had no real gift for friendship; his overbearing yet neurotic and suspicious-of-all personality would never allow him to unbend.

After Newton's election as President of the Royal Society, a position he held until his death, he reigned like a despot. His attitude even to his closest associates was at best patronizing. Halley and Newton had had their share of disagreements and squabbles, and their relationship, particularly in later years, was not without rancour. Their letters are revealing in that Newton always seems to have assumed he was Halley's superior. Even after Halley became Astronomer Royal—the most senior and respected astronomer in the land—Newton still treated him like a helper. Newton's relationship with Flamsteed was on similar footing. Both were expected to jump to it at Newton's instant command. In 1724, when Newton was pressing for a third edition of the *Principia*, he asked Halley to compute for him the place of the 1680-1 comet on a given day, assuming a parabolic orbit. This time it was Halley who committed an error, then two months later he wrote to Newton in alarm to apologize. But Newton's displeasure was not to be appeased, and Halley's recalculation never did appear in the third edition.

Halley also had another side to him. He was not always placatory before the ageing despot. At the very last meeting of the Royal Society that Newton presided over just a few days before his death he publicly scolded Halley (in his role as Astronomer Royal) for his failure to dispatch regular reports—the very offence for which Flamsteed himself had been in constant hot water. On this occasion Halley defended himself vigorously on the grounds of self-interest. He was not prepared to release observations that might allow another to anticipate his own work in trying to discover an accurate method of determining longitude at sea (using the Moon's motion) and claim the prize which had been offered by Parliament. It was an incident involving some heat on both sides, and later Thomas Hearne, anxious to make maximum capital of it, reported it as "a great Quarrel which happened between him [Newton] and Dr Halley so as they fell to bad language. This 'tis thought so much discomforted Sir Isaac as to hasten his end".

Thus, if we are to take Hearne at face value, Halley brought about not only the death of his own father but also that of the creator of the *Principia*.

After Newton's death Halley became the grand old man and sage of British science. He was never elected President of the Royal Society nor was he ever knighted. Knighthoods for pure scientists were rare bestowments in the seventeenth and eighteenth centuries. In many respects Newton's presence had overshadowed him so his achievements were eclipsed. Only in more modern times did he emerge from the shadows to be recognized as a great man of science. A year before his death he had the personal satisfaction of completing an entire 18-year observing programme of the complicated cycle of the Moon's swings about the sky. His contemporaries tell us he never lost his sense of humour. His medicine containing bitter quinine was always referred to wryly as his "chocolate".

In 1741, when he was eighty-five, his health began to fail. The end came on 14 June 1742. After savouring a glass of wine, allegedly against doctors' orders, he expired in his chair without so much as a groan.

In his later years comets were never far from his thoughts, particularly the one he had seen in 1682. In his *Tabulae Astronomicae*, published posthumously in 1749 (again reprinted in 1752 with an appended English translation), just ten years

50 Newton in old age.

51 Newton at eighty-three.

before the return of the 1682 comet, Halley refers back to his earlier prediction in emphatic terms: "... You see, therefore an agreement of all the elements in these three, which would be next to a miracle if they were three different Comets; or, if it was not the approach of the same comet towards the Sun and Earth in three different revolutions, in an ellipsis around them. Wherefore, if according to what we have already said, it should return again about the year 1758, candid posterity will not refuse to acknowledge that this was first discovered by an Engishman."

An Englishman may have predicted its reappearance, but as time drew nearer, this comet, and its fortunes, was to pass into the hands of the age-old rival across the Channel—the French.

52 Halley at eighty painted by Michael Dahl.

CHAPTER 12

1758 and Vindication

Although Newton's *Principia* had been greeted as a masterpiece by his own countrymen, there were those on the Continent, especially the French, who persistently rejected its tenets and maintained their faith in Cartesian vortices as the true explanation of gravity. It was the French Academy in particular who fought the most rigorous rearguard action and went to great lengths mounting mathematical competitions to try to discredit Newton's theories at any cost.

But the old diehards like Bernard De Fontenelle, President of the Academy, were fighting a losing battle against enlightened younger savants. Two in particular were Pierre De Maupertuis and Alexis-Claude Clairaut. In 1735 the Academy put Newton's theory to the test by mounting two scientific expeditions, one to Lapland and another to Peru to make measurements so that the shape of the Earth could be gauged. It was De Maupertuis, assisted by Clairaut, who led the Lapland Party. When both expeditions returned and the results were calculated, it was found the shape of the Earth was distinctly flattened at the poles. This vindicated Newton who had predicted such a fact in the *Principia* (*see* note 12).

Even so, some persisted in their disbelief of Newton's theories. Fontenelle himself proved to be the last voice in a lost cause when, in 1752, in his ninety-fifth year, he published his final treatise on Cartesian vortices. It was to be Clairaut who returned honour to France in 1758, a year after De Fontenelle died (*see* note 13).

The more progressive savants were aware that Halley, as the disciple of Newton and possibly a genius in his own right, had provided another sure test of Newtonian mechanics when, by its methods, he had predicted the return of the comet seen in 1682,

53 Bernard De Fontenelle (1657–1757) President of the French Academy, was the last of the important anti-Newton diehards.

54 P L De Maupertuis (1698–1759) was leader of the French expedition to Lapland in 1735 to test Newton's theory that the Earth was an oblate spheroid rather than elongated in the direction of the poles as believed by many in the French Academy. Ironically, although the Lapland Expedition (and another sent to Peru) returned to vindicate Newton, the errors subsequently found in their respective measures were so large that it was realized they could not really settle the question one way or the other! By the time this became known, however, Newton's ideas had eventually triumphed and been accepted by most of the French Academicians.

discovered by their own countrymen Picard and De La Hire. If this comet returned at the time appointed, it would provide the ultimate proof to hammer home the last nail into the Cartesian coffin.

Prior to 1758, Clairaut in particular had been thinking out methods to refine Halley's calculations so as to fix more exactly the time of the comet's reappearance. He was a brilliant mathematician from an early age. Born in May 1713, he soon became recognized as a true prodigy. By the age of twelve he astonished his seniors with his wonderful mathematical powers, and by fourteen he had written a treatise on curves of double curves which in August 1729 received the approbation of the Academy of Sciences. Because some Academicians could not believe that so mature a study could be the work of such a youth, a certificate had to be produced to prove the date of his birth.

In 1743, Clairaut published his mathematical study of the figure of the Earth and earned himself a formidable reputation. The prize he ultimately sought was a thorough understanding of the motion of the comet the Englishman had predicted would return.

As 1758 drew nearer, there were those who were openly sceptical of Halley's bold prediction (*see* note 14). Following Halley's example, there had been others who, not to be outdone, had made similar claims for others comets. One of these was Jacques Bernoulli (a member of a famous family of mathematicians and scientists) who announced the return of the comet of 1680-1 for precisely 17 May 1719 and confidently forecast it would appear in the constellation of Libra. When the time passed and no bright comet was confirmed, there were mutterings that mathematicians were no better than astrologers.

But Clairaut himself had no such misgivings and was convinced that Halley's researches into the history of the 1682 comet were soundly based. Because he was pressed for time to make all the tedious calculations he believed necessary, he enlisted the help of his friend the astronomer Joseph De Lalande. Then on Lalande's recommendation another helper was recruited. She was Madame Lepaute, born Nicole-Reine Étable de la Brière.

Lalande by this time was a well-known astronomer and held the appointment of Royal Astronomer to the French court. Madame Lepaute was the wife of a highly skilled clockmaker. As a girl she had been raised within the shadow of the Luxembourg Palace in sight of the observatory. Even at an early age she was renowned for her intelligence and wit. An anecdote from her childhood recounts that one day when a vain sister remarked to her: "I am the fairest in the family," Nicole is said to have countered quickly: "Maybe so, but I am the cleverest!"

She first became known to Lalande when her husband made a remarkable clock with only a single wheel. As the representative of the Academy of Sciences, Lalande went to visit him and inspect the unique mechanism. Then in 1755, Lalande and Lepaute collaborated on a treatise on clockmaking, and Mme Lepaute calculated a table for their book. Lalande was so impressed by her abilities with figures that when Clairaut revealed his plans and enlisted his aid, he in turn immediately enlisted her help. So great was the undertaking they had embarked upon that Lalande later revealed it was doubtful he and Clairaut would ever have finished in time but "for Nicole".

When they began, the task ahead of them was formidable. Halley had roughly estimated that the time of the comet arriving at its perihelion would be at least one year longer than the interval between the two preceding returns. The trio now strived for more exactness.

According to Lalande, they toiled incessantly, excepting meal-times, day and night, week after week, month after month. Unlike Halley, who had only approximated the influence of Jupiter, they now calculated the perturbing effects of Saturn as well. It was necessary to compute the distance of the comet from both disturbing planets, not only from 1682 but also back to the previous revolution in 1607—a period totalling more than 150 years. While this part was itself a laborious undertaking, it did not compare with the work of computing the disturbing force of each planet for the same period. Their task was handicapped by their ignorance of two yet undiscovered planets, Uranus and Neptune. Lalande wrote that he found the strain of work almost intolerable and it permanently undermined his health.

Interest in the anticipated event was mounting. Even before Clairaut announced their findings, French astronomers were busy with their telescopes. Voltaire tells us that during 1758, some observers did not go to bed for fear of missing the comet. Across the Channel, English mathematics and astronomy were in the doldrums. If there were to be honour, it would be all French.

On both sides of the Channel there were scoffers. In London

Alexis-Claude Clairaut (1714–65)

Madame Nicole Lepaute (1723–88)

Joseph De Lalande (1732–1807)

55 Computers for the 1758–9 return of Halley's comet.

the *Gentleman's Magazine* published verses expressing disbelief about Halley's prediction. In France, the critics through 1758 were in a mocking mood and kept asking where the elusive comet was hiding. In exasperation Clairaut was finally driven to reply: "The comet which has been expected for more than a year has become the subject of a curiosity much more lively than that which the public usually bestows upon questions of astronomy. True lovers of science desire its return because it would afford striking confirmation of a system in favour of which nearly all pheomena furnish conclusive evidence. Those on the contrary, who would like to see its philosophers embarrassed and at fault, hope it will not return, and that the discoveries of Newton and his partisans may prove to be on a level with the hypotheses which are the result of imagination. Several people of this class are already triumphing, and consider the delay of a year, which is due entirely to announcements destitute of all foundations, sufficient reasons for condemning the Newtonians... Therefore, I have undertaken to show that this delay predicted by Halley, far from invalidating the system of universal gravitation, is a necessary consequence arising from it; that it will continue yet longer and I endeavour to assign its limit."

In November 1758 the trio's work was complete. By the 14th the findings were in the hands of the French Academy. In spite of the monumental calculations, the exact time of the comet's reappearance was still necessarily uncertain. The known periods were quite unequal: from 1531 to 1607 the interval was 27,811 days; from 1607 to 1682 it was 27,352 days, which gives a difference of 459 days between perihelion passages round the Sun. During the calculations the question always in their minds was: would the new period be shorter or longer—or would it by chance remain the old value?

The results of the trio's work was a firm prediction that the perihelion passage would be delayed for 618 days, and the actual year would be 1759. Saturn would delay the comet for 100 days and Jupiter 518 days, bringing the perihelion passage approximately to the middle of the month of April 1759. Clairaut added that probably due to terms omitted (Uranus and Neptune were not discovered until 1781 and 1846 respectively) and possibly

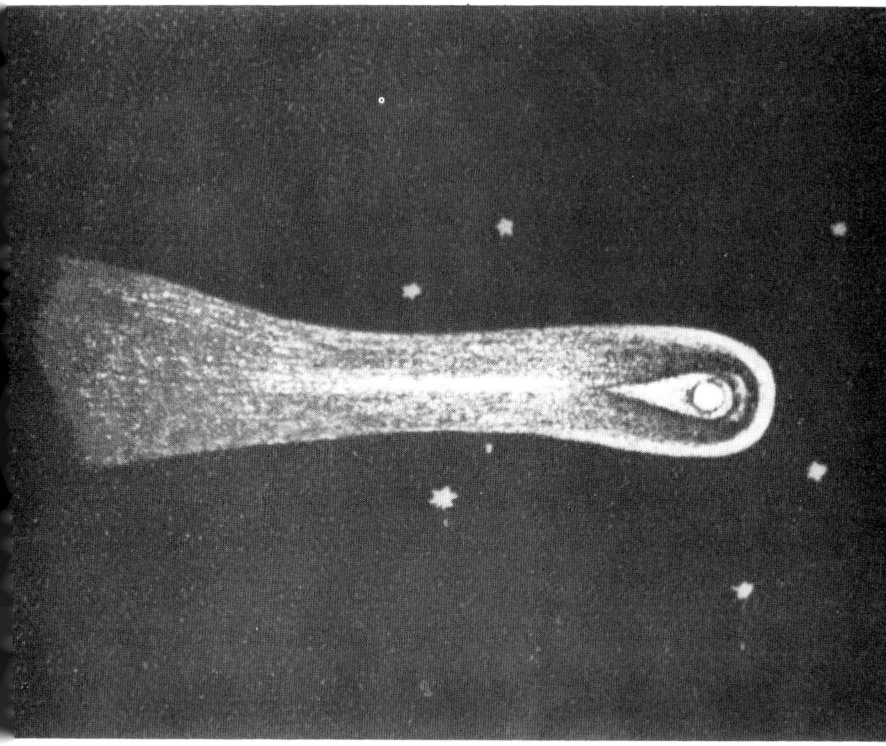

56 A stylized contemporary depiction of the appearance of Halley's comet in 1759.

errors in calculation, there might still be a difference of one month either side of the predicted date.

One eagle-eyed French astronomer who had rarely gone to bed during 1758 for fear of missing the comet was Charles Messier, a civil servant employed by the Marine Observatory in Paris. Unfortunately, in spite of his application to the task, his hopes of seeing the comet first were dashed, for there were not only Frenchmen on watch. Honour instead fell to a farmer called Johann Georg Palitzsch, an obscure amateur astronomer living at Prohlis, near Dresden, who spotted it on the evening of Christmas day 1758 through a telescope of just over 2 metres focal length.

Little is known about this Saxon farmer except he was a dedicated amateur and possessed a very keen eyesight. So acute was his vision that, according to Baron von Zach, he was able to see all four of Jupiter's brightest moons with the naked eye. During the nineteenth century a splendid water-colour of the rosy-cheeked Farmer Palitzsch used to hang in pride of place on the walls of an observatory in Prague, but this unfortunately now seems to have disappeared (*see* note 15).

After Palitzsch, next to spot it was a Dr Hoffman in Leipsic on 28 December, then there was a gap before another observer in Leipsic found it on 18 January 1759. Messier's own persistent searches for the elusive object was hampered by cloud, and he did not see it until 21 January, but after that he followed it for several nights. It was consequent to Messier's own discovery a remarkable thing happened. The Observatory Director, J N Delisle, to whom Messier was assistant, refused to allow him to disclose his discovery to other French savants. As a result, since news about the other discoveries was delayed, Messier was the only French astronomer to follow it before it disappeared into the Sun's rays on its way to perihelion on 13 March 1759.

By the end of February the comet was already too close to the Sun to be visible. It was not until the last week in March it reappeared.

On its reappearance in the morning sky it was first seen in the southern hemisphere on 26 March by de la Nux on the Isle of Bourbon (Réunion) and then in Europe by Messier on the 31st. At this time it was only visible in twilight when it hung very low near the horizon shining like a very bright star with a tail extending over 25° in length.

It was now Delisle decided to withdraw his interdict, and a formal announcement about the reappearance of the comet was made on 1 April 1759. Delisle's censorship of his assistant's earlier observations was remarkable, but we know nothing of his motivations. When Messier's earlier sightings were published, some members of the French Academy treated them as forgeries. Fortunately, in the end Messier's name was cleared, but the episode was never forgotten, and Delisle was much criticized (*see* note 16).

Clairaut, now able to fix the exact position of the comet this time round, revised his calculations and reduced the error of prediction to 19 days. Later the French astronomer Laplace showed that had Clairaut then known the true value for the mass of Saturn, the error could have been reduced to 13 days—this without taking account of several other, yet unknown, factors.

During April 1759 the comet became widely visible throughout Europe, but towards the end of the month it was difficult to follow because of its low position in the sky. It was best seen by de la Nux on the Isle of Bourbon in the southern hemisphere and by the Jesuit astronomer Father Coeurdoux at Pondicherry in India.

On 5 May its tail stretched to over 47° and it was very conspicuous. When last observed by Messier on 3 June it was still plainly visible to the naked eye, but northern observers were generally hampered by poor visibility. The comet was now retreating to the outer realms of the solar system and it was last seen by Chevalier on 22 June, observing from Lisbon.

The perihelion of the comet had taken place just 32 days before the time calculated by Clairaut and his two helpers. Their work had fully vindicated the ideas contained in Newton's *Principia* and Halley's application of them in predicting the return of the comet. The outdated Cartesian doctrine was finally dead and buried. Clairaut and his helpers had demonstrated that over an interval of 150 years of orbital motion, man could predict to within 32 days the position of a comet only visible for a short interval during its long travels in time.

As Lalande himself summed it up when he enthused over the triumph: "This universe beholds this year the most satisfactory phenomenon ever presented to us by astronomy; an event which, unique until this day, changes our doubts to certainty and our hypotheses to demonstration... M. Clairaut asked one month's grace for the theory; the month's grace was just sufficient, and the comet has appeared after a period of 586 days longer than the previous time of revolution, and thirty-two days before the time fixed; but what are thirty-two days to an interval of more than 150 years, during only one two-hundredth part of which observations were made, the comet being out of sight all the rest of the time? What are thirty days for all the other attractions of the solar system which have not been included; for all the comets, the situation and masses of which are unknown to us; for the resistance of the ethereal medium, which we are unable even to estimate, and for those quantities which of necessity have been neglected in the approximations of the calculation?... A difference of 586 days between the revolutions of the same comet, a difference produced by the disturbing action of Jupiter and Saturn, affords a more striking demonstration of the great principle of attractions than we could have dared to hope for, and places this law amongst the number of the fundamental truths of physics, the reality of which is no more possible to doubt than the existence of the bodies which produce it."

Lalande also took opportunity to express his gratitude for the help afforded by Mme Lepaute. Without her, he admitted "the work was of such drudgery to be almost impossible". But Clairaut comes off in poorer light. Although in private he too was very grateful for her assistance—referring to her as "la savante calculatrice" ("the learned computer")—and adding "her enthusiasm was boundless", he, nevertheless, suppressed a more public tribute to her in his book *Théorie des Comètes* (1760). This was to please, so the sad story goes, a woman who was jealous of the superior ability of Nicole Lepaute whom she saw as a rival; or, as Lalande himself puts it "a woman who was full of pretensions to knowledge without any sort of foundation", and a woman who "succeeded in bringing about injustice" through "a wise but weak scholar of whom she had made a conquest" (*see* note 17).

Lalande, however, made sure Nicole Lepaute earned her rightful due. When he was placed in charge of the French almanac (*the Connaissance des Temps*), she became one of his chief assistants. The pair worked together for many years. We can suspect that the astronomer was smitten and a little in love with his bright assistant, who was nine years older than himself. He tells us: "Although she was not pretty Mme Lepaute had a good share of the qualities of her sex—an elegant figure, a pretty little foot, and such a beautiful hand that M Voirot, painter to the King, after finishing her portrait asked permission to copy the hand in order to have a model of the loveliest kind to use for other paintings." We hear that her portrait took pride of place in Lalande's study; with the added compliment by him it was hung beside a very rare one of Copernicus (*see* note 18).

57 An old painting by an unknown French artist calling Halley from his grave to witness the verification that the comet of 1682 returned in the year 1758 as he predicted.

Messier, badly disappointed at not being the first to rediscover the 1682 comet and thus bring double honour to France, reaped his share of rewards in the years following. He became preoccupied by comets and made them his life's work. When Delisle retired as director of the observatory in 1761, Messier was promoted to take his place. He was a man with no inclination for mathematics or for theory—for him astronomy was the joy of observing. His first comet discovery came on the morning of 26 January 1760; and through the years he went on to make seventeen more discoveries, some of which were also found independently by others. The French became the great comet hunters of the age, and their native rivalries a legend. So successful was Messier he earned the nickname "the Ferret". Another title given him, according to Delambre, was "bird-nester" of comets—who quoted the words used earlier by Louis XV—in a eulogy on Messier in 1818. Perhaps the best-known anecdote illustrating his absolute dedication to comets records that after he had found his twelfth comet and was busy searching for a thirteenth, his wife fell ill and died. During the last few days of her life, while Messier was tending her needs and unable to observe, one of his keen rivals, Montaigne of Limoges, discovered a comet. After the funeral, when a friend offered condolences about his bereavement, Messier was overheard to remark: "Alas, Montaigne has robbed me of my thirteenth comet!" Then, suddenly realizing what answer was really expected, he quickly added: "Ah! Poor woman." A postscript to the anecdote tells us that his wife was soon forgotten, but his grieving for the lost comet continued to cause him anguish and suffering for a long time after (*see* note 19).

58 Charles Messier (1730–1817) was the great French comet hunter of the eighteenth century. Here he is seen wearing a decoration presented to him by Louis XV.

CHAPTER 13

Travellers in Time

The return of Halley's comet was a triumph and stimulated a great interest in comets generally. However, it was now retreating into distant space and would not return until 1835. Most of those who had observed the comet this time round would be dead by the time it was resighted. With the periodicity of comets vindicated, it was now opportune to watch out for others; perhaps there were some with shorter periods than 75/76 years?

Halley himself had nicknamed the 1682 comet "the Mercury among comets" in an allusion to the swiftness of the innermost planet on its journey round the Sun. By comparison the 1680-1 comet was a more leisurely traveller in time, for Halley had assigned to it a period of 575 years—inferring this was the same comet seen before in 1106, 531 and 43 BC. Comets of even shorter period than Halley's would be of great interest if they could be recognized, for astronomers would not have to wait interminably to see them come round again. While the orbits of other comets were calculated, and some early returns predicted, it was several decades later before other periodic comets were identified.

In the middle seventeenth century an interest had grown to construct lists of past comets so as to keep track of them. Hevelius had dealt with the history of past comets in his *Cometographia* in 1668, but in the western hemisphere the first proper catalogue was compiled by the Polish astronomer Stanislas Lubieniczki, who published a two-volume work *Theatrum Cometicum* printed in Amsterdam the same year Hevelius's book was published. Unfortunately the scientific content of Lubieniczki's volumes is unreliable, and many of the 415 comets included are fictitious ones.

The first truly critical study of comets was written by Nicolas Struyck who published his *Algemeene Geographie* in Amsterdam in 1740; but it was his later book *Vervolg van de Beschryving der Staatsterren*, published in 1753, which was to provide a source for the greatest comet work of all time. This was entitled *Cométographie*, compiled by the French astronomer Alexandre-Guy Pingré and published in two volumes in 1783-4. Its full title is *Cométographie; ou Traité historique et théorique des Comètes.* In spite of two centenaries since its inception, it still represents the most extant work on the subject for all aspects of cometary astronomy up to the time of its publication and is never likely to be surpassed.

But Western books are by no means the only sources for comets. Early Chinese annals, painstakingly compiled by generations of "star-clerk" civil servants, long predate those compiled by Europeans. They provide sources going back to before the time of Christendom, and in later investigations into past apparitions of Halley's comet they have proved quite indispensable (see below).

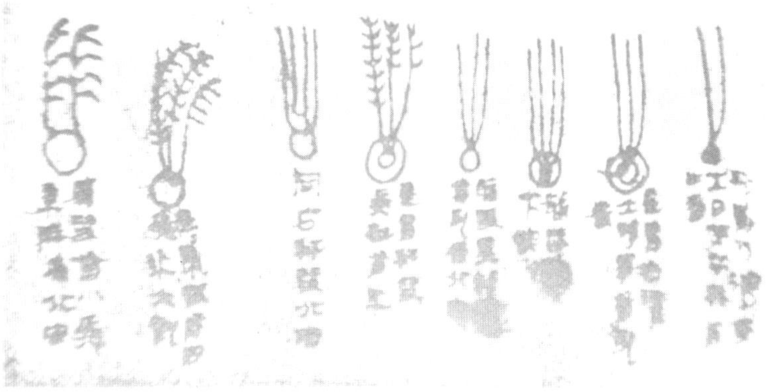

59 Early Chinese observations provide an indispensable source in tracing back Halley's and other comets. Illustrated here is a Chinese "comet atlas" dated 168 BC.

In the second half of the eighteenth century the search for new comets of short period continued. With few exceptions most of the new comets were generally dim objects seen at the very limit of naked-eye visibility or only to be identified through telescopes. One exception to the run of undistinguished comets was that seen in 1769, discovered telescopically by Messier on 8 August. In later years Messier was to remember this particular comet discovery with a special reverence for his hero—Napoleon.

With the application of telescopic power and a growing band of diligent searchers, discoveries increased in numbers over those of previous centuries. Between 1601 and 1700, thirty-five comets had been seen; between 1701 and 1800, the number more than doubled to 73, and most of these fell to comet hunters in the second half of the century. From 1801 to 1900, the number of comets identified leapt to 335...

By the second half of the eighteenth century comets had been recognized for many years as cosmic bodies occurring in significant numbers. How significant in terms of mass and size no one yet knew. It was generally agreed, however, the Earth's collision with a comet might have dire consequences. Much of this thinking referred back to 1696 when William Whiston had published a book entitled *A New Theory of the Earth*. Whiston had in fact borrowed heavily on Halley's earlier ideas and in his book elaborated on the theme that all past geological upheavals recorded in the scriptures were the results of the Earth's encounter with a comet. The comet in question was none other than that of 1680-1 (see above).

Halley claimed that the 1680-1 comet was likely the same as the one seen in 1106, 531 and 43 BC. Whiston, using Halley's alleged period of 575 years, extended its apparitions back to 2344 BC and 2919 BC. The two later dates were significant as they had been fixed by Church chronologists as choice dates for the Mosaic deluge.

Once Whiston had embarked on his historical-pseudo-scientific romance, his imagination knew no bounds. Even the Earth itself was identified as an ancient comet, but the nub of his theory was that the 1680-1 comet was dispatched by God at a previous apparition "to inflict an awful punishment on man for his sins". The result was that, at God's command, the tail of this "prodigious" comet wrapped itself round the Earth and caused the oceans to sweep away and drown the planet's guilty inhabitants "in a glorious religious purge".

For a time Whiston's book was given a scientific accolade. In public fame it temporarily eclipsed Newton's *Principia.* It was quoted in the style of exaggerated puffs common at that time as "the noblest production of genius and science that had ever been given the world". It spawned a whole series of catastrophe books written in similar vein which continue unabated to this day.

Whiston's book created a furore, even in France. Others subsequently jumped on the band-wagon, citing comets not only as portents for disasters but as actual physical causes for them. The catastrophe idea fostered by Whiston was still firmly engrained in the public's mind when in 1773 a rumour swept Paris and then all France that a comet was about to strike the Earth. Had not the illustrious Academician Joseph Lalande, of Halley's comet fame, just announced this!

When Lalande, now a celebrity, was told of what had happened, he was astonished. What had started as an innocent announcement by him concerning a paper he was to present before the Academy, on 21 April, had clearly been misunderstood. This paper, entitled: "Reflections on the Approach of a Comet to the Earth", had convinced the public and the media that a new comet was shortly to cut across the Earth's track and collide with it. Indeed the firm dates fixed for the end of the world were 20 or 21 May.

Ironically, he had not even read the paper at the Academy's meeting on 21 April for lack of time. So great was the panic in the capital that he decided he must publish a disclaimer, choosing to print it in the *Gazette de France* where it appeared in the edition dated 7 May. Unfortunately it was not emphatic enough. It did not immediately allay fears; if anything, it confirmed to the public there was some kind of an official cover-up and the distinguished astronomer only wished to disguise the terrible truth. Lalande came in for much abuse, much in the same vein as UFO buffs accuse governments of massive cover-ups today.

Voltaire himself makes jest of the affair in his *Lettre sur la prétendue Comète!* In it he first alludes to the earlier unfulfilled pronouncement of Bernoulli who had expressly predicted the

60 Joseph Lalande portrayed about the time he was embarrassed by the dramatic rumour, which swept Paris, of his supposed prediction that a comet was about to collide with the Earth.

61 William Whiston (1667–1752), who succeeded Newton at Cambridge, is now best remembered for his dramatic theories about the 1680–1 comet which he claimed had created the Deluge in a former apparition. He is seen here demonstrating the orbit of the comet.

famous comet of 1680 would return with a terrible crash on 17 May 1719... Voltaire remarks, teasingly, that perhaps Bernoulli's mistake was only one of timing and broadly hints, apropos the supposed Lalande comet, his error was "no more than fifty-four years and three days!"

Lalande's paper was finally published in the *Comptes rendus* later in 1773. Its content, sufficient to spark off the earlier panic, was innocuous enough, for it consisted principally of a table of figures which set out the distances of the crossing nodes of 61 comets.

The Lalande episode was by no means an isolated case of a public comet panic. Later there were several echoes of 1773. In 1816, again in France, rumours were rife about another approaching comet, and the end of the world was fixed as 18 July. It provided the wits with an opportunity to make jest at the expense of Lalande, now dead for the past nine years.

Three years before the scheduled reappearance of Halley's comet in 1835, another returning comet was calculated to cross the Earth's orbit, and the day of doom was fixed as 29 October 1832. In attempts to defuse the alarm raised by the rumours, F J D Arago, recalling the episodes of 1773 and 1816, drew up a notice to act as a disclaimer.

True, confirmed Arago, there was a comet due, and it was scheduled to pass close to the Earth's orbit *but not closer than 80 million km*. At the same time Arago, a fine mathematician, took opportunity to educate the public about the consequences and chances of the Earth encountering a comet, claiming the probabilities for such an event were 281 million to one. Less reassuring for the public's ease of mind was his codicil to the effect *if* a comet were to strike the Earth with its nucleus, it would annihilate the whole of the human race.

Arago's rather chilling forecast was probably and overstatement. In the light of more recent knowledge we can say it would much depend on what kind of comet hit us. Before Arago's time, the Earth most certainly collided with comets many times over; *since* Arago's time, probably at least once (*see* note 20).

62 In 1857 there was another comet alarm, and this scaremongering pamphlet sold on street corners in London for sixpence—or eight pence if the purchaser wished for a chart of its predicted track (*see* also Note 20 and Fig153).

LONDON, 1857.

JAMES GILBERT, 49, PATERNOSTER ROW.

PRICE SIXPENCE.

63 A cartoon which appeared in a French newspaper in 1857 when it was predicted a comet would collide with the Earth (*see* also note 20).

64 Comet watcher François Arago (1786–1853) was one of the giants in French nineteenth-century astronomy.

One of the most interesting comets to appear in the second half of the eighteenth century was one discovered by Messier on 14 June 1770. On 1 July it approached very close to the Earth (to within 1,200,000 km—which in astronomical terms is considered very close) and appeared to observers over $2\frac{1}{2}°$ in diameter—or five times the apparent diameter of the full Moon (*see* note 21). It was immediately suspected to be a comet of very short period, but Pingré's attempt to compute an orbit met with no success. Various orbits were computed, but the first satisfactory one was published by Lexell in St Petersburg and this indicated the comet had a remarkable short period of just over $5\frac{1}{2}$ years. Here, indeed, was a comet of the kind astronomers had long been waiting for. Unfortunately Lexell's computations were not completed until 1778, two years after the comet should have been resighted.

Lexell's study showed that earlier, in May 1767, the comet had made a very close encounter with Jupiter and this had the effect of transforming its orbit into one that led to its discovery by Messier in 1770. Lexell predicted it would be seen again in August 1781, but the comet was never found. However, he had already pointed out that in August 1779 the comet was due to make yet another very close encounter with Jupiter. He tried to take account of this but was unsuccessful. During the nineteenth century others attempted to clear up the mystery of the comet which had since become known as Lexell's comet—named after its orbit computer rather than Messier, its discoverer. Probably the best investigation of it was completed by Urbain Jean Joseph Le Verrier—the Newton of his age in French astronomy. The result of Le Verrier's researches showed that when the comet approached Jupiter in August 1779, it was then deflected into an orbit which probably rendered it invisible from the Earth at the time of its perihelion. As far as is known, the comet has never been sighted again.

During the early part of the nineteenth century the brightest comet to appear was that of 1811 discovered by Flaugergues of Viviers near Montélimar on 25 March. During the autumn it remained visible throughout the night with the result it was the most widely viewed comet in Europe before or since. Its brilliant apparition, with a tail at least 70° long, seems to have sparked the

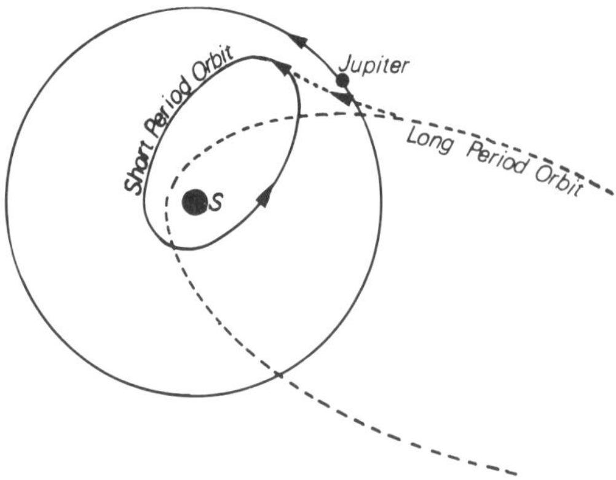

65 Because of its size, Jupiter plays a significant role in capturing long- and indeterminate-period comets and transforming them into short-period comets.

lay public's imagination. In a letter in the *Gentleman's Magazine* the following year, a correspondent claimed it had influenced the ripening of fruits such as "figs, melons and wall fruit", and "very few wasps appeared" while "flies became blind and disappeared in the season..." Yet another effect, it was claimed, was the number of females who produced twins that year.

However, the most memorable supposed effect of this comet was on the wine harvest. Long after it had passed into oblivion, London vintners' catalogues still carried adverts for the famous comet wines of the 1811 vintage, and there were references to them even at the end of the nineteenth century.

The same comet inspired terror in the Russian peasantry, for Napoleon, conscious he had been born under the influence of the comet discovered by Messier in 1769, saw its presence as a good omen for his Eastern campaign. Messier himself had actually published a book in 1808 to uphold the idea of Napoleon's birth comet, and this had the title of: *La grande Comète qui apparut à la naissance de Napoléon le Grand.* While the 1811 comet apparently had been good for wine, it had a souring influence on the fortunes of the Little Emperor, and those who recalled it in 1812 then saw it as the heralder of his downfall.

As time crept on towards 1835 and the next predicted return of Halley's comet, astronomers still sought examples of the elusive

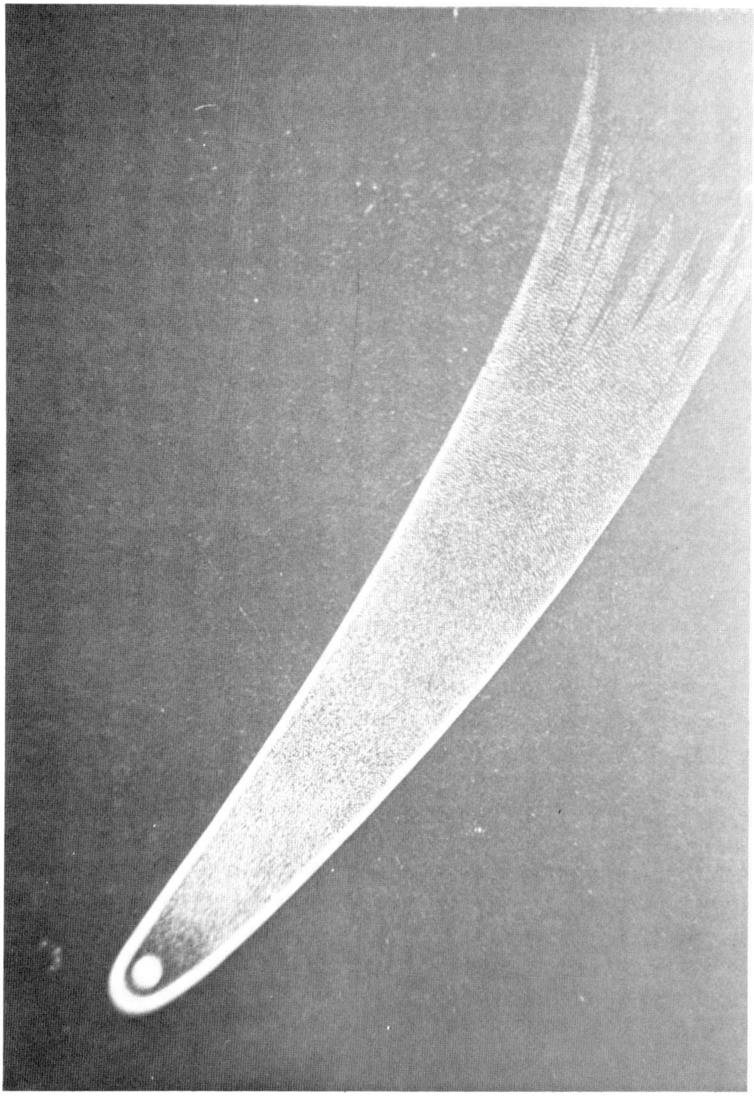

66 The great comet of 1811.

short-periodic comets like that of Lexell's which had become lost.

It was now the era of the indefatigable French astronomer Jean Louis Pons, the man who replaced Messier as the premier comet hunter in the world. His reign proved to be the most successful ever known. His record of 36 comets has never been beaten nor

67 A portrait of Charles Messi
in late middle age. His patronag
as a leading astronomer of Fran
began from the time of the *ancie*
régimes and then continued thr
the Revolution and Empire—fr
Louis XV through Napoleon I t
Louis XVIII—and then to the
beginnings of the Industrial Er
Of all those he'd known his
personal hero was Bonaparte.

68 A nineteenth-century cartoon illustrating the supposed effects of the comet of 1811 including its influence on the wine vintage that year.

69 As with the advent of the earlier Plague and Fire comets, the comet of 1811 caused a great stir and consternation among the common London citizenry as is recalled in this famous depiction of Thomas Rowlandson.

approached (*see* note 22). His only possible rivals for the future are the man-made, infra-red, astronomical satellites (IRAS), the first of which began its remarkable career as an automated comet spotter in 1983.

A comet discovered by Pons on 26 November 1818 was to herald a new comet of exceptional short period. It remained visible for nearly seven weeks which allowed time for a long series of observations to be made of it. At this time it was not an object that raised much immediate interest for it was not particularly bright, being hardly visible to the naked eye as a smudgy faint star.

In due course the observations were examined by Johann Franz Encke, who had just completed his work on researching the orbit of the 1680-1 comet. At this time Encke was at the start of a long illustrious career which finally led him to the directorship of the observatory in Berlin, the post he held when Neptune was discovered there in 1846.

At first the observations of Pons's new comet puzzled Encke. He decided to use the method invented by his old tutor, J K F Gauss, to calculate its orbit. He was soon aware that Pons's comet was a very unusual one. If his computations were correct, this was a *true* Mercury among comets, for its orbit took it round the Sun in the very short time of $3\frac{1}{3}$ years.

Now alerted to the fact that such a fast-revolving comet must often have been seen before, he searched back through the records. He was playing the game of comet detective, like Halley generations before him, and looking for its characteristic telltale orbital "fingerprints".

He was immediately struck by the fact the orbit had a resemblance to comets seen in 1786, 1795 and 1805. Further calculations soon confirmed this. Next, and notwithstanding the earlier toils of Halley, then Clairaut, Lalande and Mme Lepaute, he embarked on what turned out to be one of the most laborious and intensive hand calculations ever performed in the history of celestial mechanics. It meant computing the effects of planetary perturbations, step by step, for each previous apparition of the comet, and for six weeks he laboured alone day and night almost without stop. He realized if he succeeded, fame would be his. As a result of his intensive effort he was able to predict the comet would reappear at its perihelion on 24 May 1822, having on this occasion been delayed by Jupiter by nine days.

Encke had been aware that others were also in the race to solve the strange riddle of Pons's comet. Almost at the same time the French astronomer Arago came to a similar result, as did the Germany amateur astronomer Dr H W M Olbers. However, neither of Encke's competitors had extended the work as rigorously as he had done to give it a sure mathematical proof, and subsequently prime credit is usually attributed to Encke.

The successful recovery of Pons's comet (later renamed Encke's comet) soon after this was quickly recognized as the second great milestone in the history of cometary astronomy (*see* note 23). Another comet besides Halley's had been observed exactly where it was predicted to be. What was even more interesting, it proved there were comets in the solar system with short periods comparable with those of the inner planets.

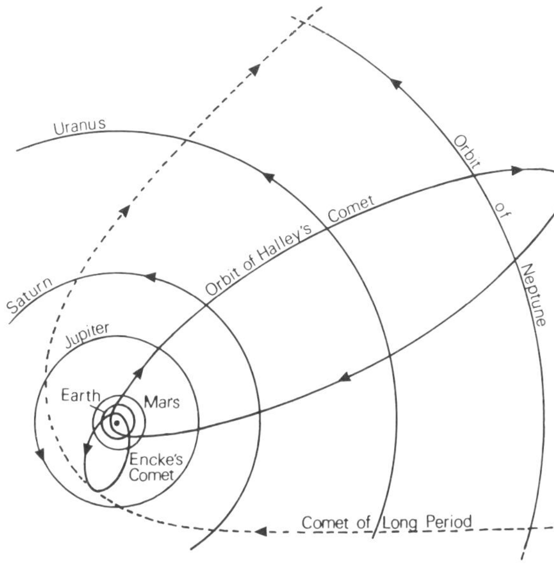

70 The orbits of Halley's and Encke's comets. Also shown is a typical comet (*pecked line*) with a large perihelion distance and a period of revolution round the Sun measured in some thousands of years.

The recognition of this new periodic comet was to raise some pertinent speculations about comets in general during the following years. Encke kept up his study of it because some aspects of its behaviour still puzzled him. The period of its revolution round the Sun seemed to be getting shorter, not by much but by an amount still just detectable by his computing methods. In 1838 he announced the results of his findings. After taking into account every conceivable planetary perturbation, the comet had returned to its last three rendezvous with the Sun $2\frac{1}{2}$ hours too early. After Encke's death in 1865, the comet was investigated by Von Asten at the Pulkova Observatory. He noted that on its return in 1868 its change of acceleration suddenly decreased by half. Then in 1871 any acceleration had vanished (*see* note 24).

Later, other comets were found to have similar "jerky" characteristics, but since Encke's comet comes round more frequently than any other, it has been studied more intently. During Encke's day it was suggested a resisting medium in space was responsible for these acceleration changes, but later this idea was abandoned. Nowadays these changes are directly attributable to non-gravitational forces at work on the comet, and therefore they cannot be computed using Kepler's laws. The most favoured idea to explain

71 (*a*) Johann Franz Encke (1791–1865) and (*b*) the short-periodic comet that now bears his name. The comet shows its typical photographic appearance without a tail. The background stars are trailed images as the camera telescope has been made to follow the shift of the comet.

a)

their cause is the so-called "rocket-effect", induced in the comet by spurts or emissions from the spinning nucleus. It is of interest that Halley's comet suffers similar acceleration changes which in the past has led to discrepancies in predicting exact perihelion dates. Confirmation that such a physical effect actually takes place awaits the outcome of planned space missions to Halley's and other comets (see below).

b) Encke's comet

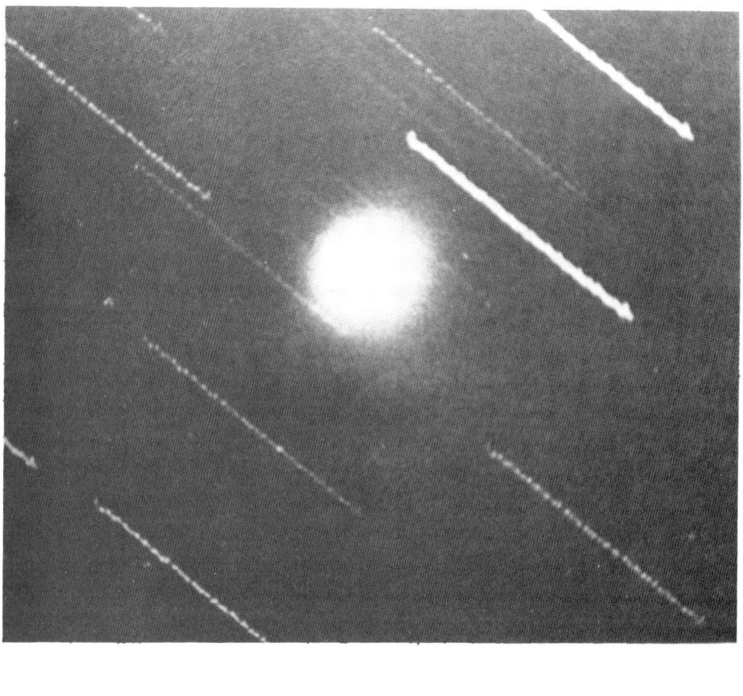

CHAPTER 14

The Return of 1835-6: The Search for Exactitude

Although celestial mechanics had undergone many refinements after 1758, the task of predicting the exact time for the next perihelion of Halley's comet in 1835 was still a formidable one. As the year drew closer, a prize worth a gold medal and 3,000 francs were offered by the Institute of France for any person who could frame up a general theory of perturbations acting on comets that could be applied in particular to Halley's comet. For a time nobody took up the challenge, and the Institute renewed its offer several times. Mathematicians with the necessary skill were only too aware of the total dedication a task like this required—not discounting a possible ruination to health and upsets in domestic routine any such declaration of intent would incur. The backbreaking toils of Halley, Clairaut, Lalande, Mme Lepaute and then Encke had long passed into folklore.

As early as 1817 a prize was also offered by the Turin Academy for a study of the comet's perturbations since 1759. This challenge was taken up by the French Baron Damoiseau, and his Essay was published in 1820 in the *Memoirs of the Turin Academy.* By this time allowances could be made for the effect on the comet by the massive planet Uranus which, since the comet's last apparition, had been discovered by William Herschel in 1781.

The upshot was that Damoiseau fixed the comet's new perihelion passage for 20 hours Paris time 4 November 1835. Consequent to this, another Frenchman, the Count de Pontécoulant, made an independent calculation and reached almost the same result except he fixed his date for perihelion at 17 hours Paris time 12 November 1835—a little over a week later.

Neither Damoiseau nor Pontécoulant, however, had completed a definitive study of the comet's perturbations by all relevant planets. This more exacting and time-consuming challenge was taken up by the German astronomer Otto August Rosenberger, who had once served as assistant at the Königsberg Observatory under Friedrich Wilhelm Bessel—the astronomer who, by one of those quirks of fate, just missed discovering the planet Neptune (*see* note 25).

In undertaking his researches into Halley's comet, Rosenberger believed it was not only necessary to account for the movements of the comet since 1759 but also to go back to follow it from 1682 to 1758. In doing so he took into account all the perturbing effects of the massive planets Jupiter, Saturn and Uranus then also those of the inner terrestrial planets Venus, Earth and Mars. Neither did he overlook the peculiar effects of the then believed resisting medium which was suspected to be operating on Encke's comet (see above). Rosenberger himself was never convinced about the reality of such a medium and therefore decided to publish two predictions: one taking account of it and another ignoring it.

Rosenberger had the advantage over his earlier rivals of having the use of a series of previously unpublished observations made by James Bradley during 1759. Bradley was the man who had succeeded Halley as Astronomer Royal. These earlier observations passed into Bessel's possession via an astronomer called Stephen Peter Rigaud. Bessel in turn passed them on to his old protégé, Rosenberger.

Rosenberger published his results in the *Astronomische Nachrichten*, the most influential astronomical journal of the nineteenth century. With no allowance for the resisting medium, the time was fixed for 0 hours Paris time on 11 November 1837; but allowing for the existence of a medium (which would have the effect of shortening the comet's period), the date would fall about a week earlier, at precisely 19 hours Paris time on 3 November.

In his paper, Rosenberger detailed the perturbing effect of each individual planet on the comet. In the case of the Earth this amounted to $15\frac{2}{3}$ days, Venus $5\frac{1}{3}$ days and Mercury and Mars together 1 day. Added up they equalled a total of about 22 days and this, it was anticipated, would hasten the comet in its orbit by that amount.

Another German computer, Lehmann, believed there was still room for a fourth voice in the matter. Although his study was not as comprehensive as that of his countryman Rosenberger, he extended his work on the orbit back to 1607 when Kepler had observed the comet. From this he concluded that the perihelion would take place on 26 November 1835, about two weeks later than the dates fixed by Pontécoulant's and Rosenberger's first choice of dates and three weeks later than Damoiseau's prediction and Rosenberger's second choice.

The search for the comet began in 1834, at least a year before predicted time of perihelion. Olbers suggested there might be a strong possibility of spotting it between December 1834 and April 1835 when its path lay across the constellations of Auriga and Taurus, regions of the sky very favourably placed at this time of year for observers in the northern hemisphere. Waiting in readiness in the southern hemisphere, at the Cape of Good Hope, was John Herschel, equipped with a 20-foot-focal-length reflector—then the largest telescope in the world. This famous British astronomer had gone to South Africa in 1834 to extend the catalogue of objects observed by him and his father, Sir William Herschel. Because of the great light-gathering power of this large telescope, many expected that Herschel would have a decided advantage over all others in the race to be first to spot the comet.

But early searches proved unfruitful, and the comet continued to elude the large band of professional and amateur astronomers anxious to stamp their names in the annals of fame as Farmer Palitzsch had done on the night of Christmas Day 1758.

One of the observatories taking part in the search was the Collegio Romano at the Vatican. It was equipped with one of the finest 15-cm refracting telescopes in the world, and its location in the clear skies of the Mediterranean gave its observers an edge over their more northern rivals who had to contend with long cloud-outs, sometimes lasting weeks on end. On the night of 5 August 1835 the Director of the Collegio Romano, Father Dumouchel, was putting in his stint at the telescope. Suddenly his attention quickened, for not far away from the position predicted by Rosenberger's ephemeris, near the star Zeta Tauri, was a faint, slightly misty, almost star-like image. Applying a higher power to the telescope, he then examined it more closely. Quite soon he concluded the object was the returning comet. Nevertheless, it was still very faint, and the position he measured for it did not please him. He identified it again on the following night, but the first observation he used for the purposes of announcing its exact position to the rest of the world was the one he made on 7 August.

During the following days the presence of bright moonlight and cloudy skies prevented independent discoveries elsewhere. It was not until 21 August it was picked up in another of Europe's finest telescopes at Dorpat in Russia by Friedrich Georg Wilhelm Struve, then Director of the observatory there. During the following week it was seen by all the leading observatories in Europe.

Struve's measures with the magnificent 24-cm refractor at Dorpat showed the comet only differed from Rosenberger's predicted ephemeris by 7′ of arc in Right Ascension and 17′ in Declination. This small difference in position indicated that the perihelion passage would be delayed until 10.30 hours Paris time 16 November—or five days later than the date predicted by Rosenberger's choice of estimate disallowing for a resisting medium and only four days later than the inspired, but less thoroughly researched, prediction made by Pontécoulant. It was Rosenberger who received the major accolade and he was later awarded the Gold Medal of the Royal Astronomical Society in 1837 for his exacting work.

During September 1835 the comet increased in brightness. By 23 September, Struve could just see it with the naked eye. A few days later it generally became visible without optical assistance.

As early as 24 September a rudimentary tail was noticed and during October it grew to a length of 20° to 25°, others reported it 30° long. The comet now became a conspicuous object as it moved through the constellations of Ursa Major, Hercules and Ophiuchus.

In spite of his large telescope and arrays of lesser instruments, Herschel, ensconced at the Cape and out of touch with the northern hemisphere, did not spot the comet until 28 October—and only then after having been notified of it by another resident British astronomer. Trouble was, Herschel lacked an exact ephemeris for the comet's path, having left England before Rosenberger's became available. Yet, in spite of this, he must have been annoyed with himself for not seeing the comet earlier, for when he did, it was already a prominent object. He later remarked: "I am sure that I must have often swept with a night-glass over the very spot where it stood in the morning sunrise; and never was surprise greater than mine at seeing it riding high in the sky, broadly visible to the naked eye, when pointed out to me by a note from Mr Maclear who saw it with no less amazement on the 24th."

A few weeks later Herschel's 20-foot telescope came into its own. Using its full aperture, he was to make some of the best detailed examinations of a comet's head ever obtained.

Towards perihelion time the comet sank lower and lower until it disappeared below the SW horizon. Last view of it was on 30 November. Then the comet remained hidden by the glare of the Sun until 31 December.

Since the 1758-9 return of Halley's comet, the number of professional and amateur astronomers had increased manyfold. So too had the power and effectiveness of telescopes. There were also many more observatories, and those like Dorpat and the Collegio Romano were equipped with primary and auxiliary instruments of very exacting qualities.

With the successful repeat prediction of the 1835 apparition of Halley's comet, the problems concerning the dynamical paths of comets appeared to have been solved. All that seemed required to gain more exactitude was application of effort combined with the necessary mathematical skills and insights. But what of the comets themselves? What were they made of? How did these wanderers into far space differ in physical make-up from the other, apparently more solid, nearer bodies in the solar system?

No one knew. Neither was it yet known for sure whether comets shone by their own light or whether they were rendered visible by reflected light received from the Sun. Comets were usually observed in telescopes as vague, misty objects, often with a bright star-like nucleus, and it was only near the Sun they began to develop their tails.

Before the reappearance of Halley's comet in 1835, there had been a dearth of bright comets in the years immediately preceding it. The last really bright one had appeared in 1811. Since then the physical nature of comets had begun to tax and puzzle the ingenuity of a new breed of astronomers who wished to know something beyond the dynamics of bodies moving in space.

Back in 1819 the French astronomer Arago had been the first to investigate the nature of the light of comets, using a polariscope. He chose the comet of 1819 which had suddenly appeared above the north-west horizon on 1 July and was discovered independently by several people. While this comet was visible to the naked eye, it was nothing like as brilliant as the 1811 comet. From his observations Arago concluded the comet shone by reflected light rather than by any light intrinsic to it. His work, however, needed confirmation and extending to studies of other comets. Needless to say, as 1835 drew closer, he anticipated applying his instrument to Halley's comet.

It was on 23 October 1835 he first turned his apparatus on the returned comet. The earlier results were confirmed: "The light of the comet," he wrote, "was not composed of rays having properties of direct intrinsic light; it was reflected or polarised; that is to say definitely it was light that had proceeded from the Sun."

Now at least one of the physical mysteries about comets seemed solved; comets, apart from their more eccentric orbits, were otherwise definitely akin to the planetary bodies in that, like them, they shone by reflected light. It was not, however, as simple as Arago and his contemporaries had first imagined. Several generations later, when the spectroscope was applied to comets, this straightforward picture was to become complicated.

Another of the mysteries about the physical nature of comets were the tenuous envelopes of nebulous matter often seen radiating from the regions of the coma and nucleus. Sometimes rays emerging from the leading side of the comet seemed to have been forced backwards by the action of the Sun, a fact commented on much earlier by Kepler and Halley. Drawings of very bright comets in the past had also occasionally illustrated this, particularly some by Hevelius, but some later observers believed them to be exaggerated features or distortions of truth. In the past observers had also reported that bright undulations or coruscations were often seen to pass down along the tails of comets. Now, in 1835, there was hope that Halley's comet would shed more information on the reality of these features.

After Herschel had finally picked up the comet in October 1835, he observed it assiduously on every possible occasion through his 20-foot telescope. He repeatedly described his astonishment at the physical changes that took place in the structure of the comet from hour to hour and from night to night. Jets, or hoods, of luminous material were constantly ejected from the nuclear region, and its appearance was in a constant state of flux. Typical of his observations is that of 28 October: "Within the well-defined head and somewhat eccentrically placed was seen a vividly luminous nucleus, or rather an object which I know no better way to describe than by calling it a miniature comet having a nucleus, head and a tail of its own perfectly distinct and considerably exceeding in intensity the nebular disk or envelope."

Herschel's observations were to confirm the long-suspected, very tenuous nature of cometary heads when he noted that a faint star, over which the comet's head was passing, remained quite undiminished in brightness even though it was within 40″ of the nucleus.

Earlier, during the first week in October, Struve was also observing the central regions of the head with his fine 24-cm Fraunhofer refractor. At one stage he noted the nucleus appeared like a fan-shaped flame emanating from a bright point; then it suddenly took the form of an oblong red-hot coal. By 13 October, he reported the centre of the comet looked like "a stream of fire issuing from a cannon's mouth after discharge", and the sparks were driven backwards as by a violent wind. At times the phenomenon was so vivid he believed the flame to be in motion "similar to the flickering of the Aurora Borealis". Then later a second small flame or jet was seen. By 5 November the comet's head was transformed and reshaped like a "powder horn" and arched in a remarkable way.

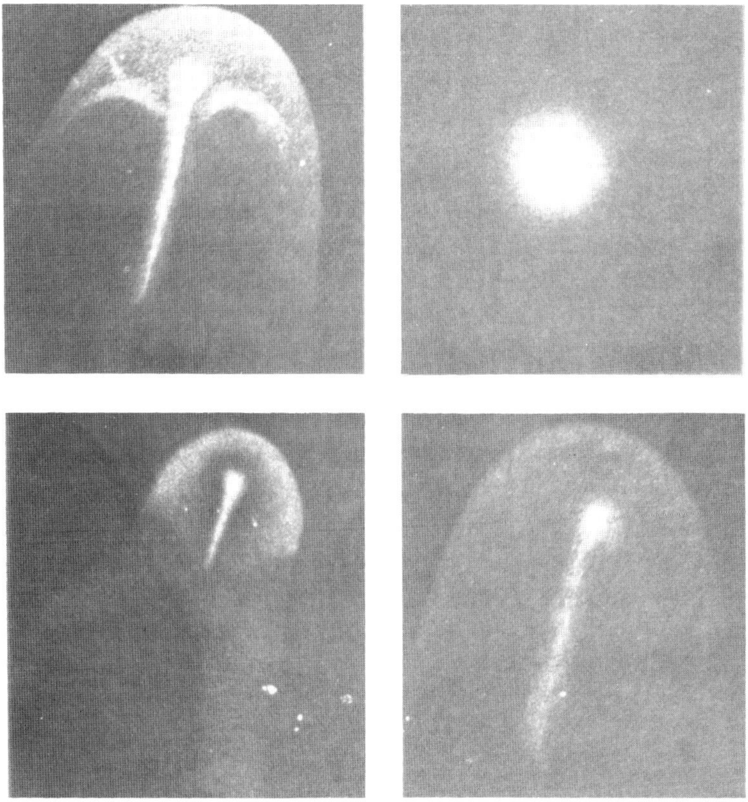

72 Drawings of Halley's comet in 1835–6 from the observations by Sir John Herschel while he was resident at the Cape of Good Hope.

73 François Arago portrayed round the time he observed Halley's comet in 1835.

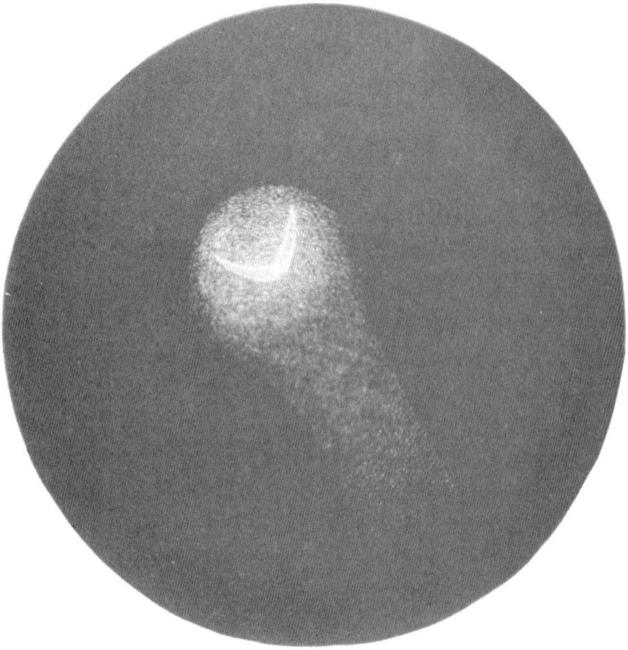

74 The telescopic appearance of Halley's comet on 11 October 1835 as drawn by Captain (later Admiral) William Smyth.

One of the father-figures of British amateur astronomy, Admiral William Smyth (then ranked Captain) described and drew much the same thing (Fig 74). On 10 October, he noted: "The comet in this evening examination presented an extraordinary phenomenon. The brush, fan or gleam of light was clearly perceptible issuing from the nucleus... On viewing this appearance, it was impossible not to recall the strange drawing of the 'luminous sector' which is given by Hevelius in his *Annus Climactericus* as the representation of Halley's comet in 1682 and which had been considered a distortion."

Other observers were also describing the remarkable activity in the head, and it was widely noted that the comet alternatively lost and regained a tail. The head would contract so it almost resembled a star, then, a few nights later, it would be swelled out into a nebulous globe which over the following nights would gradually dissolve again and finally vanish.

Bessel, observing from Königsberg, made a very famous series of drawings depicting all the changes from night to night. He was also beginning to form a theory that the activity in the tail might

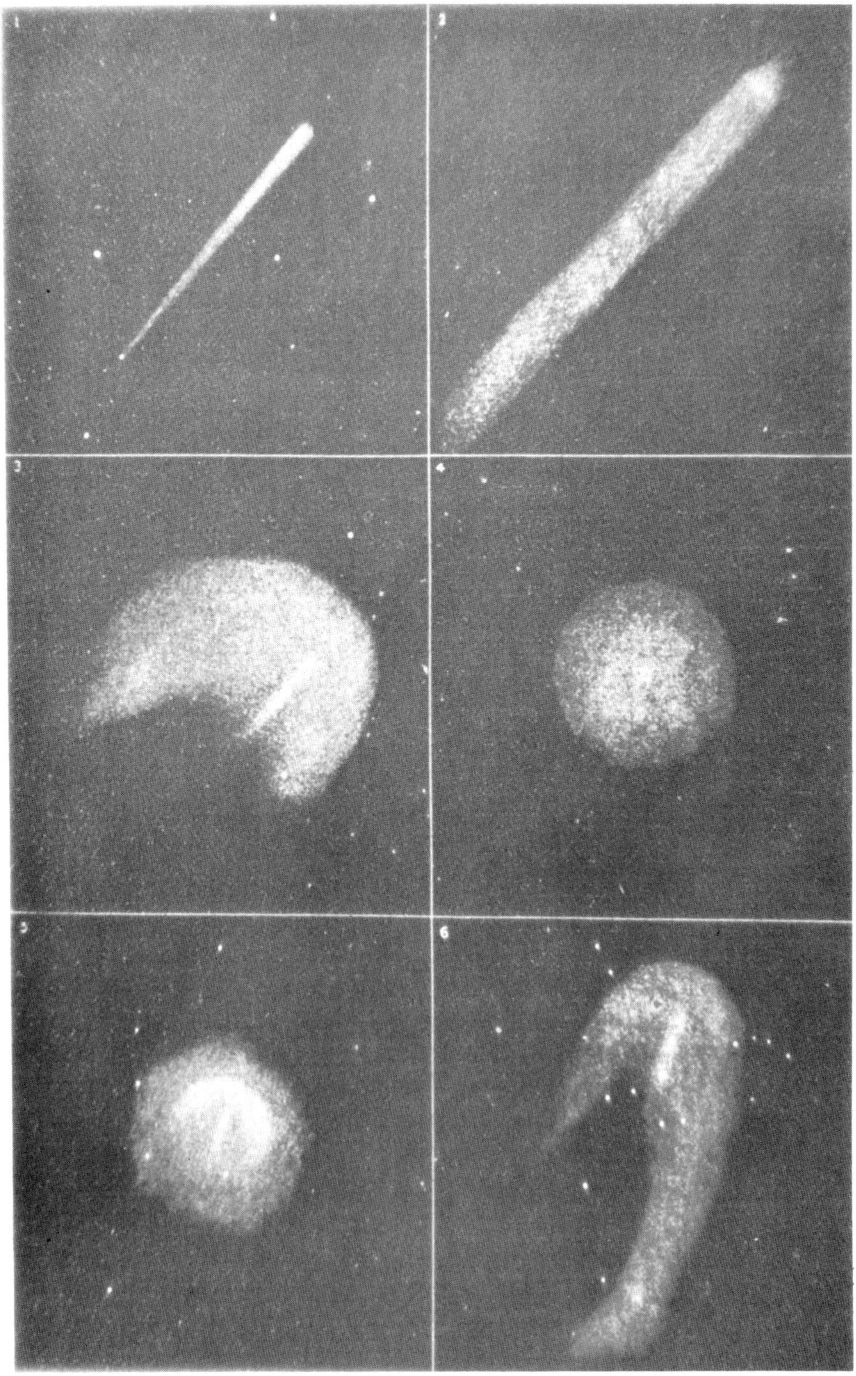

75 Drawings of Halley's comet during the 1835–6 apparition made by Sir John Herschel at the Cape of Good Hope.

76 Sir John Hersch[el]
(1792-1871).

be due to "a purely electrical phenomenon" that was operative between the Sun and the head of the comet. He was on the right track, but this puzzling physical theory how sunlight interacted with comet heads was not to be solved for more than a century until the solar wind was discovered during the 1950s.

When the comet reappeared in January 1836, after its close approach to the Sun, it remained a noteworthy sight. It now passed into the far southern skies and then began to fade as it slowly retreated in mid-February. The very last observation of it was made by John Herschel in May 1836. At this time it was very low in the sky and a dim object shining at about magnitude 9 or 10. The big telescope could not be depressed sufficiently low enough to follow it so it was observed through a portable refracting telescope of 13-cm aperture. It was this small telescope that eventually found its way to the Helwan Observatory in Egypt where it was attached to a much larger reflecting telescope to act as a finder. As he last viewed the comet through it in 1836, Herschel could not possibly have anticipated this same telescope would be in the act again when the big Helwan reflector was the first instrument to photograph the tiny image of the returning comet seventy-three years later...

CHAPTER 15

Natural History of a Comet

After Halley had identified the 1682 comet as the reappearance of those seen earlier in 1607 and 1531, other people made attempts to extend his work even further backwards. Halley himself had tentatively identified the year 1456 as another likely possibility, but he was not sure. The comets of 1531, 1607 and 1682 had given him intervals of 76 and 75 years. On this evidence alone a great comet should have appeared about the year 1456, and it just so happened that one was seen at the time the Turks were threatening to overrun Europe.

Had Halley had access to the observations of Paolo Toscanelli, the Florentine cosmographer, he would certainly have used them to include the apparition of 1456 in his schema. Toscanelli was the only European to make useful observations of the comet at this apparition when he recorded its path among the stars and described it as follows: "Its head was round and as large as the eye of an ox, and from it issued a tail fan-shaped like that of a peacock. Its tail was prodigious, for it trailed through a third of the firmament."

Toscanelli's observations were not positional observations in the modern sense but a plot of the comet's path in respect to neighbouring stars drawn on a diagram. Unfortunately this diagram remained unknown until Giovanni Celoria discovered it in the nineteenth century and was able to deduce from it an orbit for *a* comet. Celoria was unaware this could be Halley's comet, but an astronomer named Schulhof noticed it bore a very strong resemblance to the orbit of Halley's comet and its identity was soon firmly established.

Even with the 1607 apparition seen by Kepler in Prague, Halley's observational sources were limited. Indeed, it is ironic

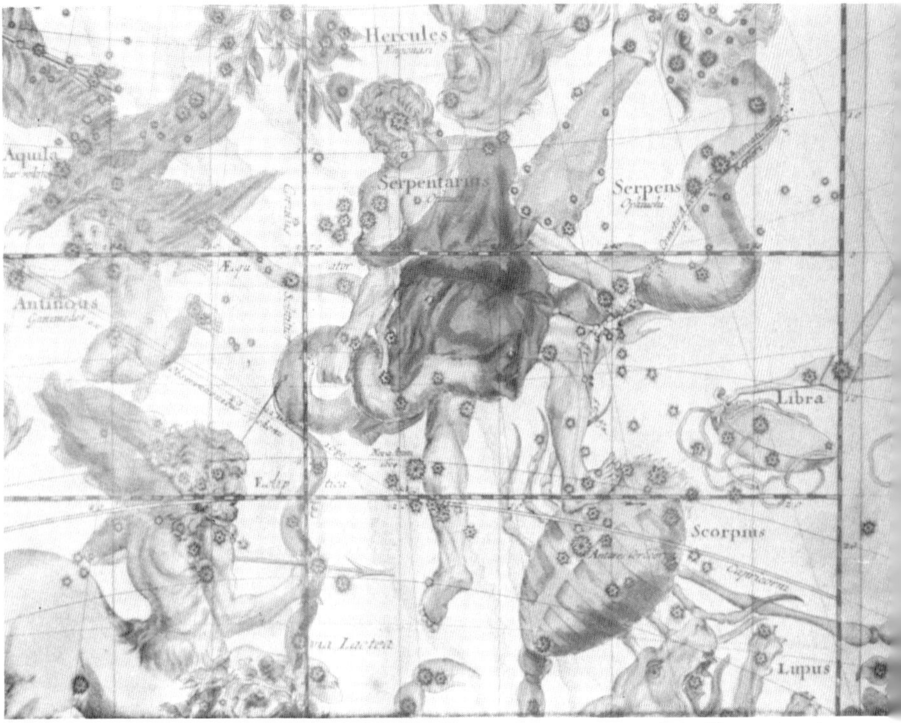

77 A very rare, much faded, late seventeenth-century star chart illustrating the tracks of the 1577 comet (according to Tycho Brahe's observations), the 1607 comet—later Halley's comet—(according to Kepler's observations) and the 1680–1 comet (according to Cassini's observations).

78 In 1456 Halley's comet was observed from 8 June to 8 July by the Florentine cosmographer Paolo Toscanelli (1397–1482). The *below* facsimile sketch is based on positions plotted by Toscanelli over an old star map depicting the constellation figures (*left*) Ophiuchus (Serpentarius), Serpens (Ophiuchi) and (*right*) Boötes. (*centre*) The semi-circle of stars belongs to Corona Borealis (Borea)—the Northern Crown.

that while Halley was acquainted with observations made in distant Prague, he was quite unaware of those made by an Englishman, Thomas Harriot, living in Sion House near Isleworth, a short carriage-ride from the City of London...

Harriot is better remembered as Sir Walter Raleigh's surveyor on his Virginia expedition; it seems it was he who subsequently introduced the potato into England and not Sir Walter, who generally takes credit for this. Harriot was a scientist of the first order, but many of his achievements were lost for generations after until fortuitously unearthed by a German baron during a visit to England (see below).

Harriot first saw the comet on 21 September 1607 when he measured its position with a cross-staff, giving the off-set distance of the comet's nucleus from various stars. He followed the comet from night to night, and his last observation is dated 13 October when the tail he describes was "obscurissima".

Halley was also unaware of the observations of the comet made by Harriot's friend and correspondent Sir William Lower in the West Country. Lower sent an account of them in a letter dated 30 September addressed "to his especiall good friend Mr. Thomas Hariotte alt Sion neere London".

Lower relates he first saw the comet at "Illford combe" (the old name for Ilfracombe). However, in another passage in the letter he implies he had embarked on a ship bound for Wales when at about midnight he saw the comet. Later he observed it again at a place called Traventi, where he had an estate through marriage. Lower also used a surveyor's cross-staff, similar to Harriot's, of the type then in vogue for measuring land. From 22 September to 6 October he observed the comet on every occasion excepting four nights when the sky was cloudy and one night when he confesses charmingly he "was a gossipinge".

Harriot's and Lower's observations remained totally unknown to their own countrymen until they fell into the hands of Baron von Zach, one of the most influential writers on astronomy round about the turn of the nineteenth century. The story relates that the Baron visited England in 1784 on a tour, and because his host knew of his profound interest in historical papers, he gave him a bundle of forgotten manuscripts belonging to Harriot, whose achievements had remained very obscure. The Baron soon realized that in his grasp was a prize indeed; these were priceless documents which shed further light on the life and works of two totally forgotten English worthies.

Unfortunately the Baron's English was poor and subsequently he made some wrong assumptions about authorship. He attributed a letter actually written by Sir William Lower to Harriot as coming from Henry Percy, of Northumberland (patron of Harriot) during the time he was held a prisoner in the Tower of London.

In due course the "deciphered" manuscripts were published by the Baron in the *Berliner Astronomisches Jahrbuch* for 1788 and later papers in the *Monatliche Correspondenz*. Fortunately the errors were spotted by Rigaud, who gave a corrected version of them in the Appendix to his now classic *Miscellaneous Works and Correspondence of the Rev James Bradley* published in 1832.

Baron von Zach's discovery in 1784 also brought to light for the first time Sir William Lower's remarkable deduction in respect to comet orbits... In 1610, after reading Kepler's *New Astronomy* and his ideas for elliptical orbits for *planets*, Lower remarked in a letter to Harriot: "...me thinks it also shews a way to the solving of the unknowne *walks of comets*". Indeed, it did exactly that—except neither Kepler, Harriot nor apparently anyone else appreciated this. It is remarkable that what everyone else overlooked was clear to this obscure Devonshire gentleman who had no pretentions to high science.

79 The path of Halley's comet in 1531 as plotted by the observations of Peter Apian in Ingolstadt and Fracastor in Verona (from *Theatrum Cometicum*).

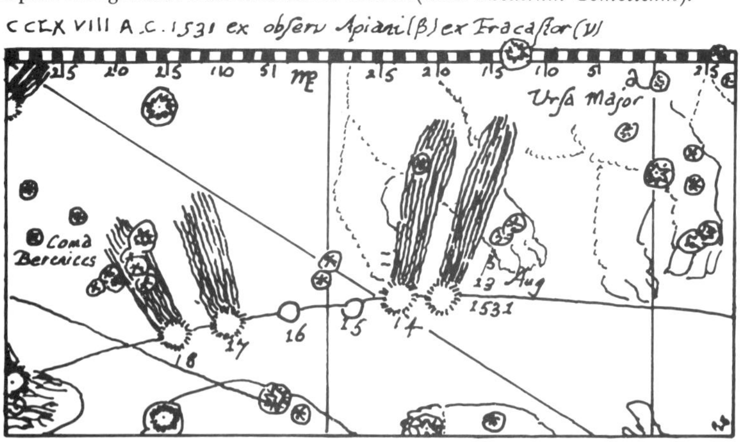

Halley's sources for the 1531 apparition were those of Peter Apian of Ingolstadt, who recorded the position of the comet on eight occasions between 13 and 23 August. Its presence near the bright star Arcturus helped him measure its position with an instrument he had especially devised called a "Meteoroscope", which he used for making graphic and rapid calculations in spherical astronomy. As a consequence of these observations Apian proved that comet tails usually point away from the Sun.

The 1456 apparition Halley had suspected but not verified was left to Pingré to convert to certainty. At this apparition the comet was described by historians of the day as "great, terrible and of an extraordinary magnitude" and "training after, its tail spread across two celestial signs 60°".

The nucleus of the comet was reported sometimes to have shone like a bright star, and the tail is described in one account as of a brilliant golden colour (attributable in the modern understanding of comets to sodium emissions triggered in the head).

In 1456 the Turks under Mahomet II were confronting the Papal forces. Both armies regarded the comet as an omen of defeat and divine displeasure. Hunniades was defending Belgrade, and so great was the terror of the comet and the Turks that

80 Halley's comet in its 1456 apparition

Callistus III ordered public prayers to be offered up for the deliverance from the comet and the enemies of Christendom. One story relates the Pope actually excommunicated the comet, but this version has been shown to be an eighteenth-century hoax—a tale put about by a French author out of humour with the Church and now thoroughly dismissed.

Going backwards from 1456, Halley had also tried to extend recognition of the the comet to 1380 and then 1305. But here he came unstuck and made errors of two and four years respectively. The actual years were later found to be 1378 and 1301. Subsequently others also jumped to wrong conclusions. In April 1835, when the resighting of the comet was keenly anticipated, an article appeared in the *Edinburgh Review* written by an author who wished to distinguish himself in historical scholarship. Primed with the (incomplete) knowledge that Halley's comet came round every 75/76 years, he looked through a catalogue of previous comets and, applying the necessary chronological yardstick, ticked off bright comets that had appeared at these intervals ... Hey presto! Nothing was simpler, and he obtained dates for previous appearances of Halley's comet in 1456, 1380, 1305, 1230, 1005, 930, 550, 399, 323, and 130 BC. We now know all but the first date, which Halley had already suspected anyway, are wrong; this would-be scholar was unaware that owing to planetary perturbations, the period of Halley's comet may vary several years either side from a median 75/76 years.

In the light of more modern records the shortest period known with certainty is 74 years $5\frac{1}{2}$ months, between 1835 and 1910, while the longest is 79 years 5 months, between 451 and 530. To arrive at sure dates for apparitions previous to 1456 required some remarkable scholarship in the hands of several men.

The first to play historical detective after Pingré was the French computer Burckhardt, who investigated the orbit of a comet observed in China in 989. This comet is also mentioned in several Anglo-Saxon chronicles. He later proved it was Halley's comet. Three other appearances were identifed by Laugier, of the Paris Observatory, who, using Chinese records, was able to identify its apparition in 451 when its reappearance coincided with the Battle of Châlons where the Roman general Aëtius defeated Attila. At this time European records which describe the comet are almost useless for scientific purposes. In these all one reads about are astrological babblings while those in the Chinese annals, recorded by the star-clerks, are cool narratives and sufficiently accurate to pinpoint the comet's path through the Pleiades and then into Leo and Virgo and to set the perihelion passage for 3 July 451. Laugier was also able to correct Halley's date of 1380 to 1378 and to identify another apparition in 760. In 1846 he published his finding in *Comptus Rendus*.

By far the most exhaustive study before the twentieth century was that undertaken by the English astronomer John Russell Hind. Almost uniquely, Hind was a skilled observer as well as a brilliant computer. He discovered three comets and ten asteroids and later became superintendent of the British *Nautical Almanac*. In his comprehensive paper about Halley's comet, which he published in the *Monthly Notices* of the Royal Astronomical Society in 1850, he brought it firmly back into the province of British astronomy from where it had languished since the death of Halley.

In his paper, Hind claimed to have identified sixteen previous returns of the comet. Four of these were subsequently called into question by later researchers; this, however, is no reflection on Hind's scholarship but indicative of the difficulty inherent in fixing sure past dates. The remaining twelve have all been vindicated. Hind only failed with the four in question because even the Chinese observations he relied upon were sometimes unclear or insufficient in number to establish the orbit with the necessary precision. His most noteworthy identification was the apparition in April 1066 when the Normans invaded England and King Harold's forces were routed. The comet was fortunately also recorded in the Chinese annals, and it was from these Hind was able to compute its orbit and identify it with the famous comet depicted on the Bayeux Tapestry.

Chinese observers first saw the comet on 2 April 1066, and Japanese astronomers independently the following morning. At this time the comet was located in Pegasus and was reported to have a thin tail pointing south-west and extending across the constellation into nearby Aquarius. Apart from the "logo" evi-

81 Perhaps the most famous of past apparitions of Halley's comet was in 1066, just before the Norman invasion of England. A representation of the comet was later woven into the Bayeux Tapestry, part of which is shown here. In the scene depicted, an astrologer is telling Harold that this strange "star" with a flimsy tail is an omen of misfortune.

82 Halley's comet in 1066 according to an illustration in *Theatrum Cometicum*. The two brightest stars (*bottom left*) are Castor and Pollux.

83 An engraving of John Russell Hind (1823–95) made shortly after 1850 when he published his definitive researches into past apparitions of Halley's comet.

84 Halley's comet depicted in *Theatrum Cometicum* as seen in AD 684 without a tail. It occasionally appeared somewhat like this during its 1835–6 apparition. The bright star in the Bull's eye is Aldebaran, and immediately adjacent (to the high right of the comet) is the Pleiades star cluster.

dence of the comet on the Bayeux Tapestry itself, only a few European observations of this comet have subsequently come to light. The most reliable one was found in 1910 in the archives of the cathedral at Viterbo in Italy which records: "In the year 1066 from the incarnation of the Lord, on 5th April a comet appeared in the East, and shone for 15 days, that is until the 19th, and the same appeared in the West in the evening on the 24th, like the eclipsed Moon, the tail of which streamed like smoke up to nearly half of the sky, and it shone until nearly the beginning of June."

The *Anglo Saxon Chronicle* records it as follows: "... In this year King Harold came from York to Westminster at Easter, which was after the mid winter in which the King [Edward the Confessor] died. Then was seen over all England such a sign in the heavens as no man ever before saw; some say it was the star Cometa, which some men call the haired star; and it first appeared on the eve of Litani-major, the 8th of the Kalends of May [24 April], and so shone all the seven nights."

According to the proverb of the time, a new "star" meant a new sovereign, and in England prayers were said to have been offered throughout the kingdom to forestall the disastrous portent. William of Malmesbury, reporting several generations after the event, told of "a certain monk of our monastery Malmesbury Abbey named Elmer, who bowing down with terror at the sight of the brilliant star exclaimed: 'Here art thou again, cause of tears to many mothers. It is a long time since I saw thee last, but I see thee now, more terrible than ever, thou threatenest my country with utter ruin'".

It is doubtful if Elmer had seen the earlier apparition of the comet in 989. It is possible, of course, as quite a number of people in history record two sightings in their lifetimes, and a twice sighting of Halley's comet is no real test of longevity. Mark Twain saw it in 1835 and then again from his death-bed in April 1910. Lewis Swift, a doyen among nineteenth-century American comet hunters, also lived during both these apparitions, but by 1910 his wonderful eyesight had failed him. If Elmer had indeed seen a comet earlier, it was probably the bright comet of 1006, reported in Europe and in China, or possibly another, for there are several round this period still uncatalogued because their paths are unknown.

85 While it is no real test of longevity to have lived through two apparitions of Halley's comet, the double event evokes significance in folk history. This stamp issued in 1985 commemorates the 150th anniversary of the birth of Mark Twain (1835) and, indirectly, the comet's return in 1910 when this American author died, shortly after resighting the comet.

86 Halley's comet over Jerusalem during its AD 66 apparition as depicted by a seventeenth-century artist.

An apparition of Halley's comet recognized by Hind was that of 1301 when it was observed in Europe and China. This apparition is supposedly recounted by Giotto's painting, the "Adoration of the Magi" where the comet is seen as a direct representative of the Star of Bethlehem (Fig 87). However, if Giotto depicted a comet he actually observed, it is more likely to have been the very bright, second comet of 1301 (Fig 87) visible, more appropriately, just before Christmas that year rather than

87 Shortly after 1301, Giotto Di Bondone painted his famous "Adoration of the Magi" in the Scrovegi Chapel, Padua, where a comet representing the Star of Bethlehem is depicted immediately above the open-sided "stable". Some have identified this as an impression of Halley's comet at its 1301 apparition. However, I dispute this identification and believe Giotto's comet is likely the *second* bright comet of 1301(*see* below) seen more appropriately just before Christmas (rather than Halley's seen earlier in September/October). It was recorded by several observers, but unfortunately its orbit is unknown and therefore does not appear in most modern catalogues.

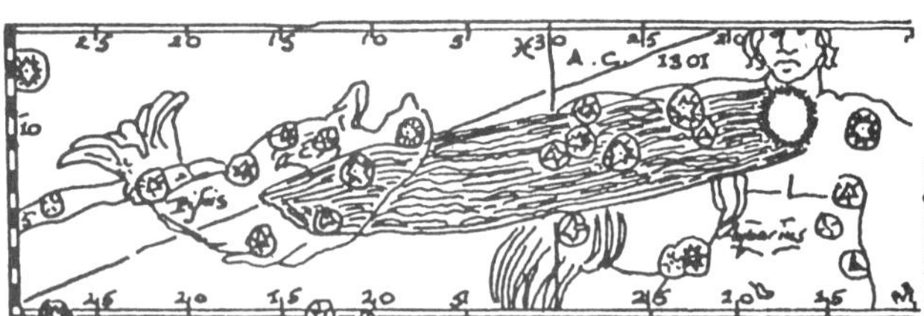

The second bright comet of 1301, based on a drawing by Stanislas Lubieniczki, c1668

Halley's comet which appeared earlier, in September and October. Two other apparitions of Halley's comet also recognized by Hind were 12(−11) BC and AD 66. Over the years these two apparitions have been involved in much speculation as possible sources for the story of the Star of Bethlehem. At the 12(−11) BC appearance, the earliest recorded by Hind, the comet is reported to have terrified the Romans before the death of Agrippa. Likewise, next time round in AD 66, it terrified the Jews when they were hard pressed by the Romans at the siege of Jerusalem when Josephus records that several prodigies announced the destruction of the city and there was "Amongst other warnings, a comet, of the kind called Xiphias, because their tails appear to represent the blade of a sword". It was also at this return the comet is alleged to have been seen by St Peter, just before his martyrdom.

One of the earliest pictures of the comet is that at the 684 apparition which appears in the *Nuremberg Chronicle*, and this certainly resembles a comet much more than the stylized logo later embroidered into the Bayeux Tapestry.

88 The appearance of Halley's comet in AD 684 as illustrated in the *Nuremberg Chronicle*.

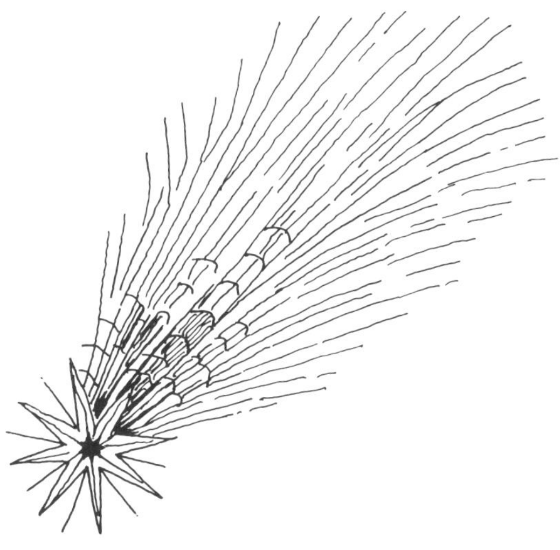

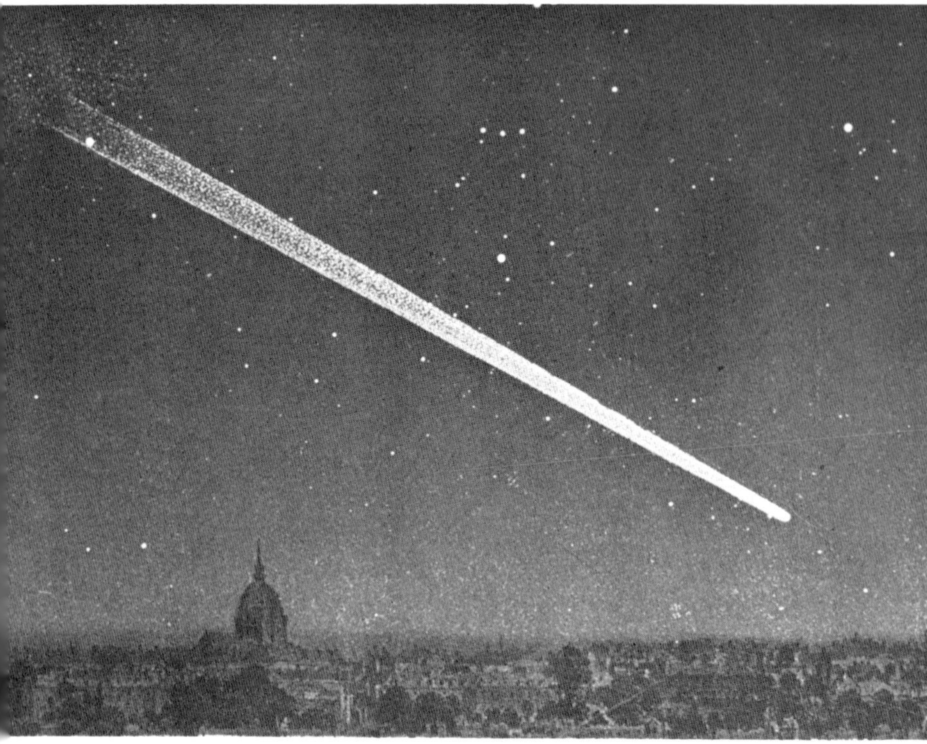

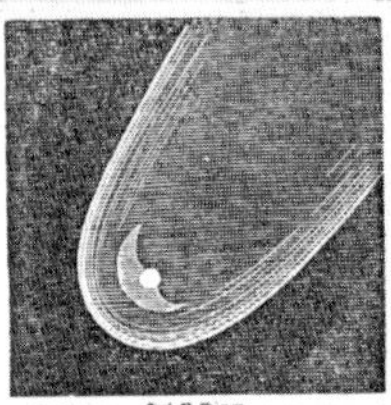

Sept. 21, 8h p.m.

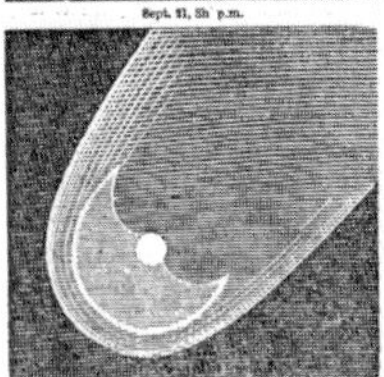

Sept. 24, 8h. p.m.

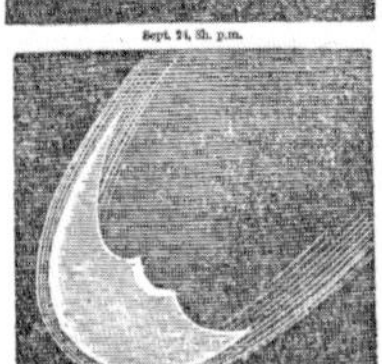

Sept. 28, 7½h p.m.

TELESCOPIC APPEARANCES OF DONATI'S COMET AS SEEN FROM THE CAMBRIDGE OBSERVATORY.

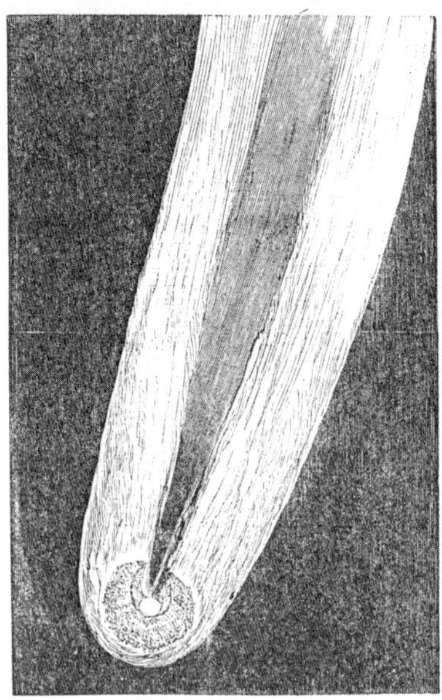

TELESCOPIC APPEARANCE OF DONATI'S COMET AS SEEN FROM SLATER'S OBSERVATORY, EUSTON-ROAD, ON THE MORNING OF OCT. 1.—SEE NEXT PAGE.

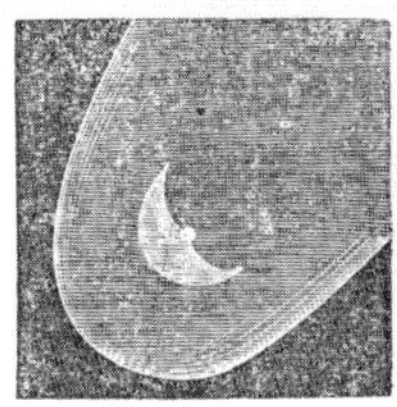

Sept. 27, 7h. p.m.

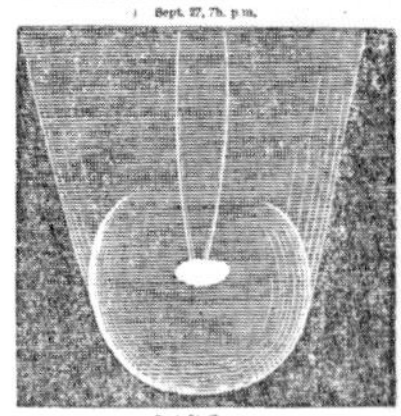

Sept. 30, 8h. p.m.

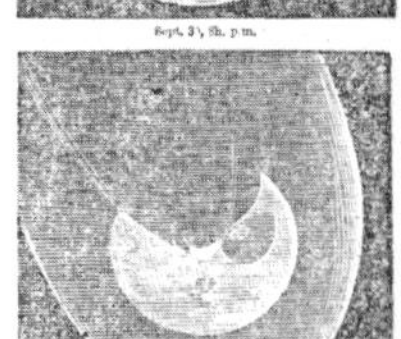

Oct. 5, 6½h. p.m

TELESCOPIC APPEARANCES OF DONATI'S COMET AS SEEN FROM THE CAMBRIDGE OBSERVATORY.

89 Two great comets of the nineteenth century: (*upper left*) The great comet of 1843 as seen from Paris. (*above*) The classic picture of Donati's comet as seen over Paris. It caused a great stir, and millions throughout the world flocked outdoors to see it. (*lower left*) The issue of the *London Illustrated News* for 22 October 1858 printed these spectacular views of its head as seen from Cambridge, England, and a Mr Slater's observatory in Euston Road near present-day King's Cross Station.

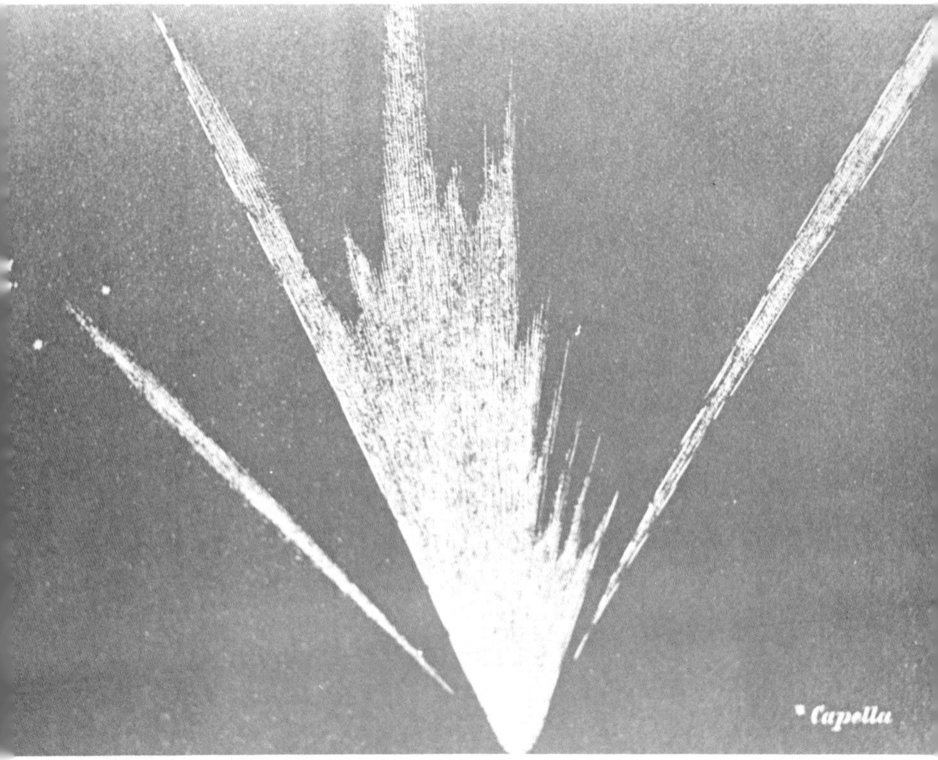

90 A naked-eye view of the great comet of 1861 as seen when the Earth passed through its tail on 30 June that year at a point two-thirds the distance from its nucleus.

CHAPTER 16

The Return of 1909-11: Through the Tail!

Flushed with his success in predicting the perihelion of Halley's comet to within four days in 1835, Pontécoulant, the French mathematician, extended his calculations forward to the next return and had already published them forty-six years before the comet was to make its twentieth-century début. According to this reckoning the comet would come back to perihelion during May 1910...

It was known that between 1835 and 1910 the comet would undergo great perturbations by Jupiter, especially in the earlier part of this interval. The question now asked was: by how much?

About the period Pontécoulant made his forecast, the Swedish astronomer Anders Jonas Ångström also became interested and, following Hind's indentifications, predicted the comet would in fact not return until some time during or even after 1912.

Then, after the turn of the century, two British professionals working at Greenwich threw down the gauntlet and, assisted by three volunteer helpers, set out on the time-consuming task of attempting to forecast the comet's next perihelion as precisely as humanly possible. With Hind's researches into its past appearances, over fifty years before, the comet had come back into the care of Halley's own countrymen. Now the five-man team of Britishers (one was an Irishman) were determined not to let it escape again into the hands of the old rivals across the Channel.

Heading the quintet were P H Cowell and the Irishman A C D Crommelin; their assistants were W M Smart, F R Cripps and T Wright. All were civil servants employed at Greenwich Observatory where Halley had reigned as incumbent Astronomer Royal for twenty-two years. The task before them was formidable. Using Hind's list as a basis, their plan was to go back to first principles and endeavour to retrace the comet step by step to each apparition, making sure that at every return it was positively identified before proceeding back in time to the next earlier one.

By now comets were recognized as very flimsy bodies. While sometimes voluminous, they, nevertheless had little mass compared with the planets or their satellites. In 1868 a comet was known to have passed very close to Jupiter and then to have wandered right through its system of satellites. Although the comet itself was thrown into an entirely new orbit, the encounter did not affect the planet nor its satellites in any way which could be measured.

The practice of computing a comet's future orbit was by now a well-established ritual. It was recognized that to obtain the most exact results, care had to be taken of *all* the different perturbations a comet may be subjected to during its wanderings in the solar system. In the case of Halley's comet its movements are chiefly determined by the Sun's attraction alone, but there are occasions when its presence near Jupiter, or one of the other large planets, creates a significant third force as Halley and Newton had first recognized back in the seventeenth century.

In the period 1836-1910 it was known by the British team, notwithstanding Ångström's prediction, that this time round the comet's revolution interval would be shortened, and, paradoxically, its quick return would be due to *delays* caused by Jupiter.

This comes about because the orbits of the comet and that of Jupiter cross approximately at two points, and the time occupied by the planet in moving from one to the other in its orbit is nearly equal to that occupied by the comet in its own orbit. As a consequence, if both are near one crossing-point together, they will also be near the other point together.

During the interval 1836-1910, Jupiter got behind the comet on both these occasions and pulled it back—the direct effect of which was *delay*. However, an indirect effect overpowered this. The comet as a result did not travel as far afield when it was checked in its path at the outset and therefore as a result came back to the Sun *more quickly*.

At its limit, Halley's comet journeys out to the boundaries of the solar system far beyond the orbit of Neptune to a distance of 4,800 million km. The peculiarity of its elliptic motion, under the gravity of the Sun, is that it spends half its revolution time dawdling in the small arc beyond Neptune's orbit. This is a consequence of Kepler's second law, viz, equal areas in equal times. While it spends *years* out at this part of its orbit, it only spends *weeks* in the inner, opposite arc as it rounds the Sun (Fig 91). At its remotest from the Sun, the lingering part of the comet's journey becomes accentuated as the time of revolution increases.

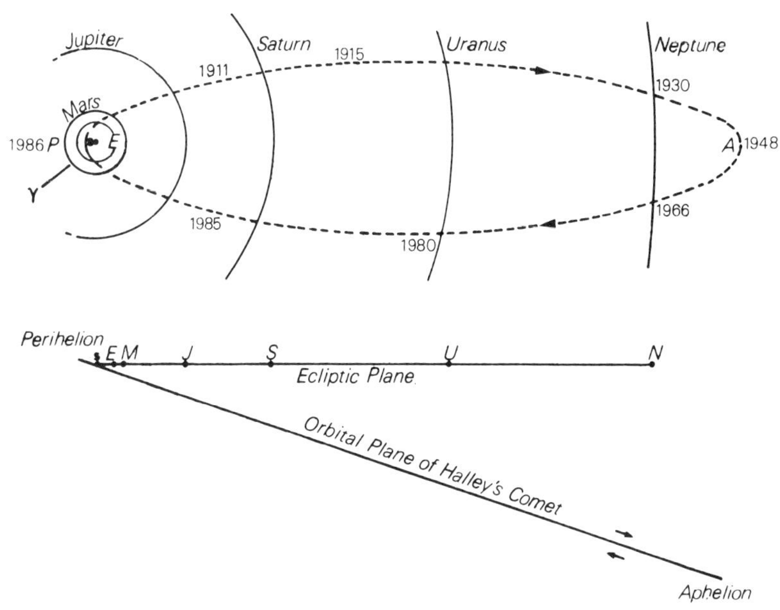

91 The orbit of Halley's comet. The dates (1911–1986) indicate the various positions of the comet during its last revolution round the Sun. Owing to Kepler's second law, the comet spends half its revolution period dawdling beyond Neptune's orbit. Unlike the major planets, whose orbits lie in one plane (the ecliptic), many comets have steeply inclined paths. Also unlike the planets some comets go round the Sun in a clockwise direction (retrograde), and Halley's comet is one example of this type.

For the 1910 apparition the Greenwich team decided to apply to the comet a new mathematical technique developed a little earlier by Cowell to investigate the orbit of Jupiter's tiny eighth satellite and which led to its rediscovery by observers. However, before any ephemeris for Halley's comet could be computed in its forthcoming apparition, the five-man team had first to face up to the laborious grind of back-tracking the comet through history.

In spite of the past researches of Pingré, Burckhardt, Laugier and Hind, plus some by oriental scholars, several previous identifications were still uncertain. It was now the observations contained in the Chinese annals proved indispensable. Hind had identified the comet in old records by tracing out its observed path across the constellations, but some of his information was often vague and at best uncertain. In contrast, the modus operandi adopted by Cowell and Crommelin was to concentrate on accurate calculations of the lengths of successive revolutions round the Sun as the comet was affected by perturbations. Hind had identified Halley's comet as a comet that had appeared in 1223 which chroniclers cited as being the instrument and precursor of the death of the French king, Philip Augustus. After due deliberation, Cowell and Crommelin discarded this comet and gave preference to another seen in August and September 1222 that probably reached perihelion in September. Chinese annals record it as having a tail "31 cubits long" and noted it "perished in two months".

Cowell and Crommelin also looked again at Hind's identification of 912. There were probably two comets that year. While Hind chose one visible in May, the Greenwich pair preferred one seen in August.

Halley's comet should also have appeared in 837. European and Chinese accounts again imply two bright comets that year, and it is possible there were three or even four. Cowell and Crommelin chose one that was at perihelion at the end of February or beginning of March. It is now known that on 9 April 837, Halley's comet approached the Earth to within 6 million km and in a single day traversed over 60° of sky. Chinese records report a tail of "80 feet", equal in modern terms to more than 100°. The comet must have presented a dramatic and frightening sight, and there is little wonder the ancients went in terror of it. Except for 1910, its tail then was the longest ever seen visually on any comet.

In 608, Hind had chosen a comet observed by the Chinese that passed over the constellation Auriga, Ursa Major and Scorpius. Cowell and Crommelin rejected this and preferred a comet seen the previous year in 607.

When the Greenwich team arrived at Hind's identification of 12 BC and duly confirmed it, they then moved back into terra incognita. The history of the comet was now very uncertain, but using the Chinese records as a guide, they identified the comet's orbital fingerprints back to 87 BC, 163 BC and 240 BC (Table 2).

TABLE 2

Returns to Perihelion of Halley's Comet

(Cowell and Crommelin)

240	(BC)	May 15?		912	(AD)	July 19	
163	(BC)	May 20?		989	(AD)	Sept. 2	(Burckhardt)
87	(BC)	Aug. 15		1066	(AD)	Mar. 25	(Hind)
12	(BC)	Oct. 8	(Hind)	1145	(AD)	April 19	(Hind)
66	(AD)	Jan. 26	(Hind)	1222	(AD)	Sept. 10	
141	(AD)	Mar. 25	(Hind)	1301	(AD)	Oct. 23	(Hind)
218	(AD)	April 6	(Hind)	1378	(AD)	Nov. 8	(Laugier)
295	(AD)	April 7	(Hind)	1456	(AD)	June 2	(Halley, Pingré)
374	(AD)	Feb. 13	(Hind)	1531	(AD)	Aug. 25	
451	(AD)	July 3	(Laugier)	1607	(AD)	Oct. 26	Halley
530	(AD)	Nov. 15	(Hind)	1682	(AD)	Sept. 14	
607	(AD)	Mar. 26		1759	(AD)	Mar. 12	
684	(AD)	Nov. 26	(Hind)	1835	(AD)	Nov. 15	
760	(AD)	June 10	(Laugier)	1910	(AD)	April 19	
837	(AD)	Feb. 25	(Hind)	1986	(AD)	Feb. 9	(*see* note 29)

This is where Cowell and Crommelin decided to call a halt; beyond this date-line any references were too conjectural. They were aware, of course, that Pingré in his *Cometographie* made references to comets in even remoter antiquity. However, the earliest that seemed to rest on reliable evidence was a comet seen in China in 611 BC. Cowell and Crommelin believed it possible that this might have been Halley's comet, but its sure identity was wanting. Yet another apparition was perhaps that of 467 BC seen in Europe and China and is perhaps the same comet mentioned by Aristotle in his book *Meteor.*

On 28 July 1908, Crommelin wrote to his friend the astronomer H H Turner, the Savilian Professor of Astronomy at Oxford: "We have now carried back Halley's comet to BC 87

The Catalogue.

"A"	"C"	Year of Discovery	Discoverer	Designation	Perihelion-Passage T		Argument ω	Asc. Node Ω	Inclination i	Perih. Dist. q	Eccentricity e	Reference
		B.C. 年		B.C. 年	月	日	°	°	°			
1	1	467	(Chinese)	467	?		109	27	162	0.60	0.97	
2	2	372	(Aristotle)	372	winter		120	270–330	under 150	Very small	1.	Pingré I
3	1	240	(Chinese)	240	May	15	109	27	162	0.60	0.967	MN 68
4	1	163		163	May	20	109	27	162	0.60	0.967	,, ,,
5	3	137	(Chinese) ?	137	Apr.	29	350	220	160	1.010	1.	Am Alm (1847)
6	1	87	(Chinese)	87	Aug.	15	109	28	162	0.60	0.967	MN 68
7	4	69	(Chinese)	69	July		150	165	70	0.79	1.	Am Alm (1847)
8	1	12	(Chinese, Europ.)	12	Oct.	8.80	108	28	162	0.60	0.967	MN 68
		A.D.		A.D.								
9	1	66	(Chinese)	66	Jan.	26	108	29	162	0.583	0.967	AN 143
10	1	141	(Chinese)	141	Mar.	25	108	29	163	0.60	0.967	MN 68
11	1	218	(Chinese)	218	Apr.	6	108	30	163	0.60	0.967	,, ,,
12	5	240	(Chinese)	240	Nov.	10.0	82	189	44	0.372	1.	MC 10
13	1	295	(Chinese)	295	Apr.	7	108	31	163	0.60	0.967	MN 68
14	1	374	(Chinese)	374	Feb.	13	108	32	163	0.60	0.967	,, ,,
15	1	451	(Chinese)	451	July	3.5	108.5	32.5	164	0.60	0.967	,, ,,
16	1	530	(Chinese)	530	Nov.	15	108	33	163	0.60	0.967	,, ,,
17	6	539	(Chinese)	539	Oct.	20.61	{255.5 / 75.5}	{58 / 238}	10	0.3412	1.	MC 2 ; 16
18	7	565	(Chinese)	565	{July / July}	9.0 / 14.5	70 / 79.5	158 / 159.5	118 / 121	{0.719 / 0.832}	1.	MC 10
19	8	568	(Chinese)	568	Aug.	29.32	24.33	294.25	4.13	0.907	1.	CR 22
20	9	574	(Chinese)	574	Apr.	7.28	15.37	128.28	46.52	0.963	1.	AN 21 ; 23
21	1	607	(Chinese)	607	Mar.	20	108	34	163	0.60	0.967	MN 68
22	1	684	(Chinese)	684	Nov.	6	108	35	163	0.60	0.967	,, ,,
23	1	760	(Chinese)	760	Jan.	11	107.5	36.5	163	0.60	0.967	,, ,,
24	10	770	(Chinese)	770	June	6.64	86.77	88.90	120.48	0.603	1.	AN 23
25	1	837	(Chinese)	837	Mar.	1	108	37	163	0.59	0.967	MN 68
26	11	868	(Jap., Chin., Eur.)	868	Mar.	4	277	305	65	0.42	1.	JJAG 10
27	1	912	(Japanese)	912	July	20	109	39	163	0.59	0.967	MN 68
28	12	962	(Chinese)	961	Dec.	30.16	81.73	355.67	95.43	0.5369	1.	Obs 20
29	1	989	(Chinese)	989	Sept.	2.0	110.67	43.15	163.0	0.585	0.967	AN 143
30	13	1006	(European)	1006	Mar.	22	93–94	38	162.5	0.583	1.	Pingré I
31	1	1066	(Chinese, Europ.)	1066	Mar.	27	105.3	38.6	163.5	0.59	0.967	MN 68
32	14	1092	(Chinese)	1092	Feb.	15.0	30.67	125.67	28.92	0.928	1.	AN 27 ; CR 26
33	15	1097	(Chinese)	1097	Sept.	21.9	125.0	207.5	73.5	0.7385	1.	MC 2 ; 16
34	16	1132	(Japanese)	1132	Aug.	30.20	114.5	201.1	106.4	0.736	1.	TA 5
35	1	1145	(Japan., Chin.)	1145	Apr.	19	105.4	39.9	163.5	0.60	0.967	MN 68
36	1	1222	(Chinese, Europ.)	1222	Sept.	10	105.5	42	163.5	0.60	0.967	,, ,,
37	17	1231	(Chinese)	1231	Jan.	30.301	121.3	13.5	6.08	0.948	1.	Pingré I
38	18	1240	(Japanese)	1240	Jan.	21.09	331.3	124.5	73.4	0.668	1.	TA 5
39	19	1264	(Chinese, Europ.)	1264	July	19.79	159.57	140.92	16.48	0.825	1.	*Hoek*
40	20	1299	(Chinese, Europ.)	1299	Mar.	31.312	103.80	107.13	111.05	0.318	1.	Pingré I

Remarks : A=Apparition, C=Comet, ( ) in Record.

92 The first folio of the comet catalogue (1936) of the Japanese astronomer A S Yamamota listing the earliest reliable observations of comets which are mostly based on Chinese observations.

Note: the comet designated 1 in the "C" column is Halley's comet. However, the apparition of 467 BC is still an uncertain identity.

(August) with certainty (one revolution earlier than Hind's list), and with fair probability to BC 240 (May). Hind is one and a half years too late for his 608 AD return (it really was 607 March), but all his earlier returns are right up to the beginning of his list (−11 = BC 12). We find 12.9 April 1910 for the next passage, but we are going over the work again by a new method."

Subsequently, Cowell and Crommelin put back the perihelion date to 16.61 April 1910.

As with the previous apparitions in 1758-9 and 1835-6, special efforts were mounted to detect the comet as early as possible. This time, however, there was a big difference from former times. The new tool of photography was available and, since the pioneer comet pictures of the 1880s, there had been rapid technical advances. No one among the professional ranks now dreamed of wasting long hours glued to the eyepiece of a telescope sweeping the skies when the photographic plate could penetrate much farther into space and then after development be examined and measured in comfort the next day.

First photographs known to have been taken in attempts to catch the comet were those of 22 December 1908 by O J Lee at the Yerkes Observatory, Chicago. Also in the race early were the observers at the Royal Greenwich Observatory, keen to back up their colleagues as often as the prevailing clouds and fogs of London would allow them to view the sky.

It was thought a unique opportunity to see the comet early would be offered during the total eclipse of the Sun in Tasmania on 8 May 1909; then, disappointingly, it was realized the comet would have set below the horizon before the eclipse began.

As the summer of 1909 approached and then passed, observers became anxious. It was the middle of September before any positive word was forthcoming. Then, flashed round the world, came the telegraphed announcement that the comet had at last been resighted by "Wolf of Heidelberg".

Maximilian (Max) Wolf, Director of the Königstuhl Observatory, was a past-master at the art of celestial photography. Earlier, on 27 December 1891, he had detected his first asteroid using a technique he pioneered called the trail method. Now, on a plate he had exposed on 11 September 1909, near the star Gamma

Gemini in a position not more than 10′ of arc from that predicted by the British team, he had picked out a minute, strange image—quickly identified as the returning comet. At this time the comet lay at a distance of 512 million km from the Earth.

After the first announcement, other observatories searched their earlier plates. Unbeknown to them, observers at the Royal Greenwich Observatory had already recorded it on 9 September, but as it was unrecognized, they had let the great prize slip from their grasp. At Helwan Observatory, in the transparent skies of Egypt, the minute image of the comet had impressed itself as early as 24 August. Even Wolf now discovered that he had overlooked an earlier image on a photograph taken on 28 August.

Following its resighting, some quick computations revealed the actual perihelion passage would be 19.64 April 1910 (later corrected to 20.18 April 1910)—or to within about three days of Cowell's and Crommelin's second, more accurate, prediction of 16.6 April 1910. In some respects this was a disappointing result for the British team after the effort expended. Nevertheless, it was a much closer prediction than any other. The Lindemann Prize of 1,000 marks which had been offered by the Astronomische Gesellschaft for the most successful prediction was now awarded Cowell and Crommelin.

The 3.03-day discrepancy greatly puzzled Cowell and Crommelin, but it convinced them there were other unknown forces at work on the comet. On 5 March 1910 they wrote: "It now appears from observation that the predicted time of 16.61 is 3.03 [3.58] days too early. At least two days of this error must be attributed to causes other than errors in calculation or errors in the adopted position and masses of the planets."

Crommelin's personal copy of their winning Essay which he received and annotated with his name on 11 April 1910 is now a prized possession of the author (Fig 97). Other annotations by Crommelin on its inside pages indicate that years later he was still having ideas how their old prediction might yet be improved to match the actual perihelion time more closely. (*see* note 29)

After the reappearance of Halley's comet had been announced in the autumn of 1909, telescopes, large and small, were trained on to the rare visitor. The first visual sighting, as opposed to an

93 Max Wolf (1863–1932), Director of the Köningstuhl Observatory, Heidelberg, was the first to announce the recovery of Halley's comet on 11 September 1909. He photographed it using a 72-cm telescope (*illustrated*) when the comet was still 3.4 A U distance from the Sun (in the asteroid belt) and shining at about magnitude 16.

94 In 1909 an intensive photographic search was made for Halley's comet using this 75-cm reflector at the Greenwich Observatory. Although tiny images of the comet were recorded on 9 September, two days before Max Wolf of Heidelberg announced its recovery, the proximity of the Moon discouraged the Greenwich observers from a thorough inspection of their photographs, and the faint images were not recognized until the recovery telegram was received from Heidelberg.

95 The 76-cm reflector of the Helwan Observatory in Egypt. Unknown to observers using it, it was the first telescope to record the tiny image of Halley's comet when it returned in 1909. It was also the telescope to last record the retreating comet in June 1911. The small finder seen above the main telescope was previously used by Sir John Herschel to view the comet in South Africa during the 1835–6 apparition.

96 Four photographs taken shortly after Halley's comet was recovered in September 1909 showing its shifting position (arrowed) against the background stars.

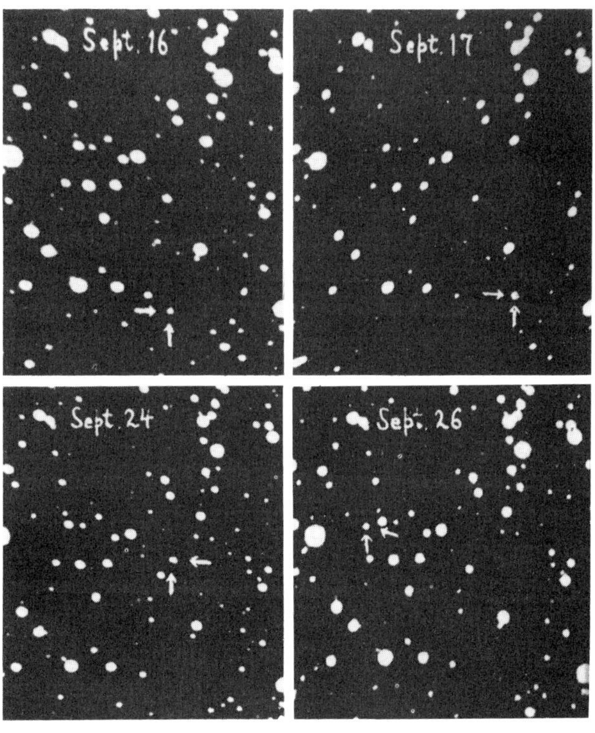

97 A C D Crommelin (1865–1939) and (*left*) the title-page for the prize-winning *Essay* for the return of Halley's comet in 1910. This is Crommelin's personal copy and is annotated in his hand with his name and the date of receipt (11 April 1910). Annotations (*right*) by Crommelin inside his copy of the prize-winning *Essay*. This and other pages retain his later recalculations in attempts to reconcile the 3-day gap between the predicted time and the actual time for perihelion.

image on a photographic plate, was reported on 15 September 1909 by the eagle-eyed American, double-star observer, S W Burnham, using the 101-cm Yerkes Observatory refractor then, and still, the largest telescope of its type in the world.

During September, and through the remaining months of 1909, the comet gradually brightened as it approached the Sun. On 19 October, E E Barnard, a colleague of Burnham, estimated it was 13^m. By November it was 12^m, then in the middle of the month a rudimentary tail was first detected. By late November it was $10^m.5$, and one observer reported fluctuations in brightness.

By January 1910 the comet was getting brighter by the day. Then overnight, when all interest was centred on Halley's comet, a much more brilliant interloper suddenly appeared and for several days captured the attention world-wide.

This unexpected comet first revealed itself in the southern hemisphere in the dawn skies of 12 January when diamond miners and railway workers in South Africa independently discovered it as they travelled home after working nightshift. A few days later it was an even more spectacular object and on the 17th it could be seen in full daylight just 4° from the Sun's edge. Towards the end of January the tail had grown to a length of 40°, and the comet had begun to move northwards. It now had a distinct yellowish tinge to it owing to sodium emissions released in the head of the comet after its close approach to the Sun. At its maximum brightness it even outshone Venus, but the brilliant display did not last for long. By middle February the comet had diminished to below naked-eye visibility and by this time was a very unspectacular object. It continued to fade, and by 9 July was at the limit of visibility even with the largest telescope.

During January 1910, astronomers had been watching both comets. By the end of January, Halley's comet had brightened to $9^m.3$; in February it was 8^m, and the tail was 40′ long. The head was now showing a system of rays, and small jets were seen emerging from the nuclear region in a way reminiscent of the 1835-6 apparition, except on a smaller scale.

By the beginning of February it was plainly visible in amateur telescopes, shining like a small nebula. On 9 February, Wolf claimed to have glimpsed it with the naked eye. Several observers were now reporting a multiple nucleus.

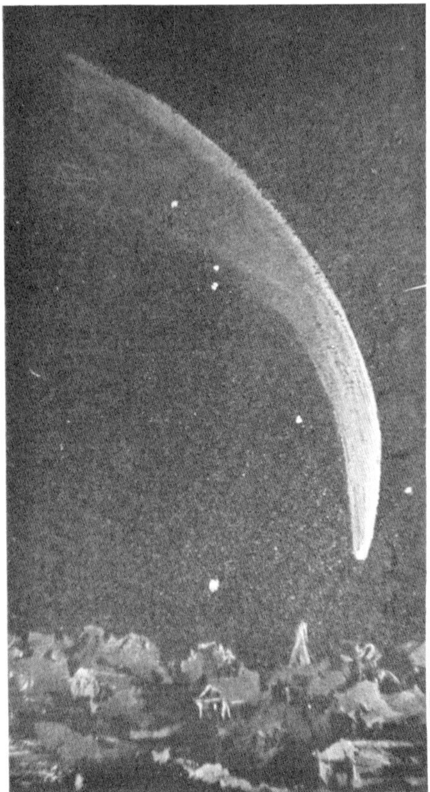

98 Two naked-eye impressions of th
Daylight comet in January 1910,

(*a*) as seen from the North of Englan

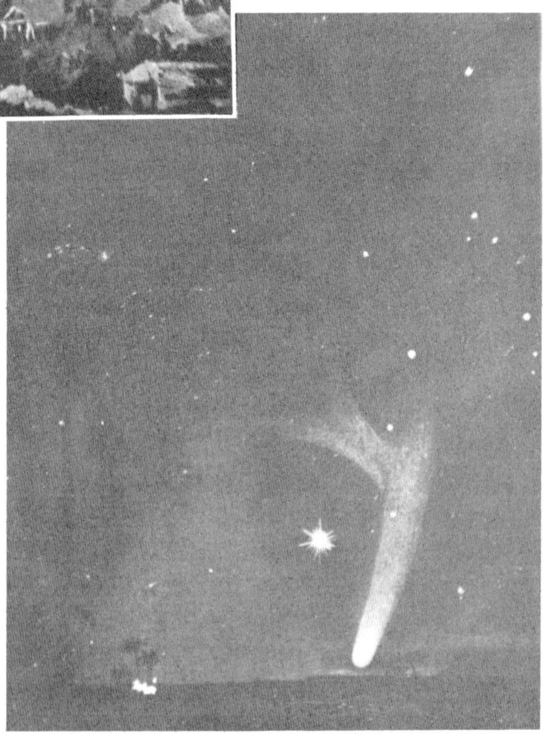

(*b*) as seen from North Africa.

In early March its brightness was still officially around 8^m, and photographs now plainly showed the tail and the numerous ray structures, and jets emerging from the head.

In the latter part of March, as it neared the Sun, it passed out of the night sky and remained invisible until the middle of April. When it reappeared, the comet was close to perihelion and had brightened to $2\overset{m}{.}5$. Photographs taken at this time show parabolic envelopes of fine tenuous material and some denser concentrations emerging from the head.

By early May the tail was 6° in length and divided. By 17 May the tail had lengthened enormously; but opinion and estimates of its length at this time are conflicting. Some observers were tracing out the tail to over 70° and estimated its width as 9°, others much less than this.

In the tropics especially, the comet was a glorious early-morning spectacle. It was now close to Venus, and this pair of brilliant objects in near conjunction presented a rare sight, stirring even the lethargic from their beds to witness it. By 1910, astronomers, like other scientists, had slipped into the habit of using the unemotional jargon of their trade. In hundreds of reports on Halley's comet we can read about coma size, magnitude, length of tail in degrees, etc, but few let themselves go as their less inhibited predecessors once did. Purple-passage descriptions were now frowned upon as old-fashioned and "unscientific". In contrast, laymen had no such inhibitions. On 18 May a London correspondent based in Accra, West Africa, wrote home: "... Here everyone has gone mad over it, and we all get up at 4 am, and sit and gaze at it till it gets light. It is *the* most wonderful thing ever seen. The comet itself rises far above the horizon, but its tail which stands straight up, is like the rays of a very powerful search-light—so long, that it reaches from the horizon to the very roof of the heavens; and so broad that it occupies roughly one fourth of the arc of the sky; and its light is so powerful that combined with Venus (which is also lovely just now), it has almost the effect of a midnight sun. The natives are frightened to death of it, and will have it that it means an earthquake coming..."

But geographical location was all important, and the spectacle

99 Halley's comet photographed by E E Barnard 4 May 1910.

100 Halley's comet as photographed in the Helwan reflector 2 June 1910.

101 Halley's comet juxtaposed with Venus in May 1910. Note the disruptions along the comet's tail.

102 In May 1910 the magazine *Punch* published this cartoon to greet Halley's comet juxtaposed with its barbed comment about amateurism versus paid-for professionalism as allegedly personified by Louis Bleriot, the French aviator who the year before had been first to fly the English Channel.

103 Even in 1910 advertisers were already aware of the benefits of using highly topical subject-matter in their advertising promotions.

from the north of England, in very unfavourable latitudes, gave rise to disparagement. An astronomical correspondent there wrote: "My first view of this truly pathetic body was on the evening of Sunday, May 22nd, 1910. When I pointed it out to some friends thus: 'There you see the celebrated Halley's Comet, which for more than two thousand years has been wheeling about the sun, and which...' my remarks were unceremoniously cut short by the entire party declaring that they refused to believe it. On my assuring them that it was so, the look of disappointment was general, and thenceforward none but myself found further interest in the insignificant, elusive, nebulous smudge which masqueraded under the guise of the great 'Halley's'.

"Next evening, when the sun-glow had become subdued, out again timidly came the comet-phantom, and hundreds of watchers in my vicinity saw it and disbelieved. Neither would they be comforted. As the evening went on, these watchers grew not merely discouraged, but undisguisedly disgusted. Their desire by that time was not so much to see a magnificent comet as to meet those people who had so airily promised them one."

A friend of this correspondent, more favourably placed in South Africa, later commiserated with him in a letter, adding: "... here the comet and Venus together was a truly superb spectacle that enraptured me. Their combined light was so brilliant that reading a newspaper was quite easy."

Elsewhere in the world cranks were predicting that the "blazing star" signalled Armageddon, and the credulous stared at it in stupefied horror much like their forebears round the time of the Plague and the Great Fire once did. In the American West, itinerant patent medicine vendors were doing a brisk trade selling comet pills to ward off its evil influences and cashing in on the heaven-sent opportunity to fleece the ever gullible public. Early in February 1910 the press reported that a neurotic in Valencia, Spain, became obsessed with the idea that the comet would collide with the Earth on 19 February and, deciding there was no hope for mankind, plunged to his death from a bedroom window, killing himself instantly. It was shades of Lalande and the Paris doom-watchers of 1772... Then when King Edward VII suddenly died in London on 6 May 1910, there were those who saw it as a manifestation of the malign influence of the comet now adorning the morning sky.

By the middle of May 1910, as the London correspondent in Accra had noted, the tail had lengthened enormously. On 18 May the comet's orbit carried it between the Earth and the Sun. Because of this unique event, an American expedition was speedily sent out to Hawaii to observe the Sun's disk at the critical time; but no trace of the comet's nucleus projected on the bright disk was seen. From these observations it was concluded that had there been a solid body as large as 320 km in diameter, forming a nucleus to the comet, it would have been readily identified.

On 19 May tail streamers from the comet were distinctly visible in spite of strong moonlight, and the "train" of the comet stretched across the sky for at least 140°, even surpassing the tail-length reported by the Chinese in the AD 837 apparition. As in AD 837 the comet was now close to the Earth. On 20-21 May, when the Earth was due to pass through the tail, a special look-out was kept to note any peculiar atmospheric or meteorological phenomena which might be directly attributed to the influence of the comet. However, the Earth only passed through the outer edge of the tail and at its closest was at least 8 million km distant from the comet's head. Owing to the brilliant moonlight, no

104 In May 1910 the Earth passed through the tail of Halley's comet. The figure shows how the comet (which moves contrary to the Earth's motion) made its nearest approach on 18 May.

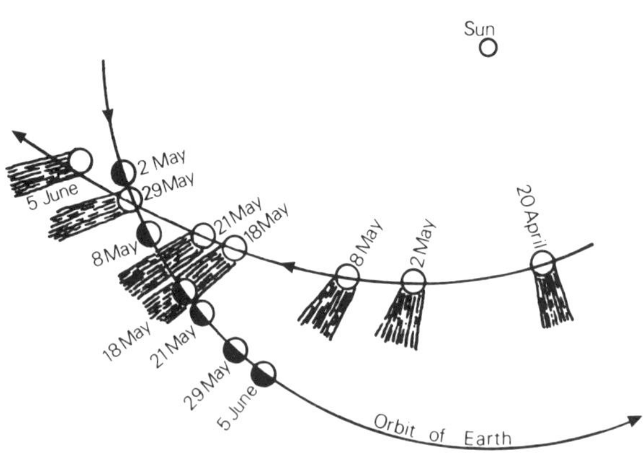

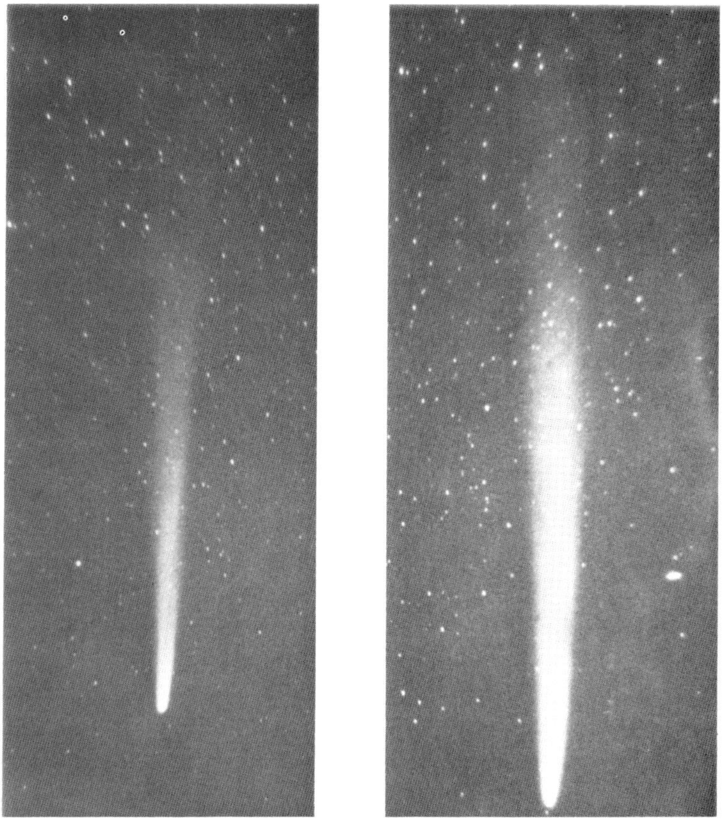

105 Halley's comet on 12 and 15 May 1910 photographed by the Ellerman Expedition to Honolulu. This expedition had gone to the island to watch when the comet transitted the Sun, but no indication of the comet's dark nucleus was seen.

positive atmospheric effects were ever confirmed; but, as anticipated, because of the wide publicity afforded the event in the world's press, there were subsequently hundreds of eye-witness reports of unusual happenings which allegedly resulted from it. (*see* note 26).

By the end of May the comet was still a 2^{m} object, but the brilliant display was now coming to an end. As the comet gradually drew away from the Sun, it faded rapidly. At the beginning of June it had dimmed to $3.^{m}5$; in July $5.^{m}5$, and by the end of the month $7.^{m}0$.

On 12 November 1910, now well on course towards its aphelion beyond Neptune, Barnard estimated the comet's brightness as 11^{m}. In April 1911 it was still in view, and at the end of the month, Barnard was reporting it to be of variable brightness and shining around $14^{\mathrm{m}}.5$ to 15^{m}. Last view of it visually was on 29 April at the Helwan Observatory, and the same observatory followed it photographically until June 1911 by which time it had receded from the Sun to a distance of 837 million km, to a point well beyond the orbit of Jupiter.

106 The changing appearance of Halley's comet from 26 April to 11 June 1910, photographed from Honolulu.

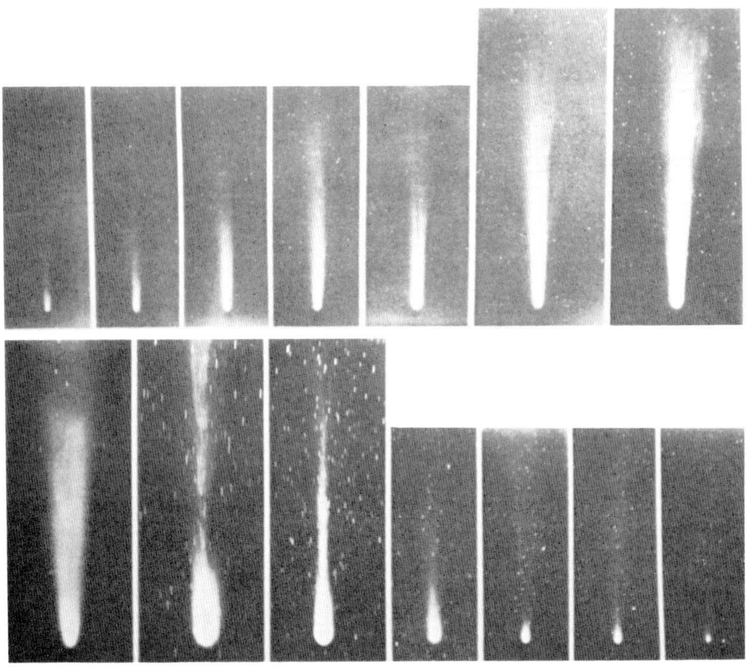

CHAPTER 17

The Nature of Comets

Since the appearance of Halley's comet in 1835 and Arago's confirmation that it shone by reflected light derived from the Sun, other advances had taken place in understanding the physical nature of comets. In 1864, Giovanni Donati (discoverer of a brilliant comet in 1858) applied a spectroscope to Tempel's comet and, analysing its light, was able to identify the bright Swan bands due to the presence of carbon. In due course other comets were examined over the years, and gradually a picture about the various constituents to be found in them slowly began to emerge, but this was not without confusion, which still exists to some degree today.

One surprise feature was the discovery that while Arago's observations were true, this was only part of the story, for the presence of bright emission bands indicated that part of the light of a comet was triggered off by sunlight interacting chemically in some mysterious way on material within the head.

The advent of photography to comets also helped increase knowledge about their physical structures. Some attempts to photograph comets during the late 1850s and early 1860s were failures, but with the invention of new plates in the 1880s, some fine pictures of late nineteenth-century comets were obtained. Photographs had the advantage of providing a permanent record which could be studied at more leisure from the comfort of an office during daylight hours and they showed details, especially in comet tails, unseen visually at the telescope.

The first photographic discovery of a comet was made accidentally during a total eclipse of the Sun in Egypt on 17 May 1882 (*see* note 27). Soon after 1892, when Barnard made his own first discovery of a comet using photography, the practice

107 The first photographic discovery of a comet was the Sungrazer of 1882, caught on this plate exposed to record a total eclipse of the Sun in Egypt.

108 The Great Daylight comet of September 1882. This photograph was a milestone in cometary astronomy, for it was the first really successful picture taken of a comet.

109 Morehouse's comet (1908 III) photographed with the Greenwich reflector on 29 September 1908. This comet exhibited some strange behaviour, radically altering its features from night to night, and it provided the first evidence that comets contain poisonous cyanogen gas later detected in Halley's comet in 1910 (*see* Fig 110).

became a routine activity, for in this way much fainter comets were found than could be seen visually.

How comet tails were formed remained much of a mystery. Morehouse's comet, discovered in 1908 shortly before Halley's comet reappeared, provided several surprises. One was the identification of poisonous cyanogen gas in its spectrum hitherto unrecognized in comets. More immediate, dramatic surprises were in the tail itself, which rapidly changed from night to night. Arthur Eddington, later Sir Arthur and famous for his cosmological ideas of the expanding universe, was then Chief Assistant at the Royal Greenwich Observatory and managed to measure the velocity of material ejected from the comet's head along the tail. He was astounded to find it travelled at a speed of 112,000 km per hour—or expressed another way, it represented a solar repulsion force 800 times more than the pull of gravity.

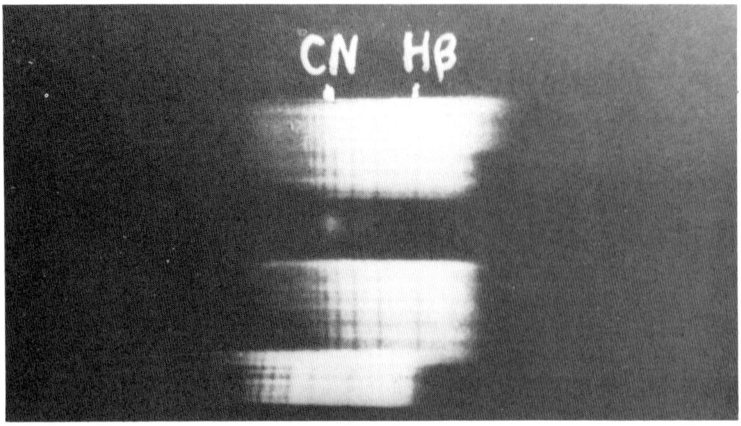

110 A spectrum of Halley's comet taken on 14 January 1910 indicating the presence of cyanogen (CN) and hydrogen (Hβ).

During the 1909-11 apparition of Halley's comet, its tail was studied intensively. What now became certain was that comet tails were composed of very fine tenuous material. In 1910, K Schwarzschild estimated that if the tail of Halley's comet at its greatest length were composed of only fine dust particles, it might weigh possibly 1 million tonnes, but if it were only tenuous gas, not more than 1,000 tonnes. A gas tail of this type is much less dense than the Earth's atmospheric envelope, therefore no effects would be noted when the Earth passed through it as was indeed found to be the case in May 1910.

The realization that a comet's tail was a very tenuous structure, as proven by Halley's comet, gave rise to the popular, if slightly disparaging, description of a comet as being the nearest thing to "a great big bag full of nothing". While this is a tongue-in-cheek exaggeration, it, nevertheless, contains an element of truth, and space-age discoveries in the late 1960s and 70s were to confirm the idea. By the first decade of the twentieth century it was clear that apart from the "solid" nucleus of a comet, of which absolutely nothing was known, comets as cosmic bodies were lightweights. Indeed the density of a typical comet envelope (excluding the nucleus) might be appreciated by imagining a volume of space measuring one cubic km with twelve small marbles spaced equidistantly inside it.

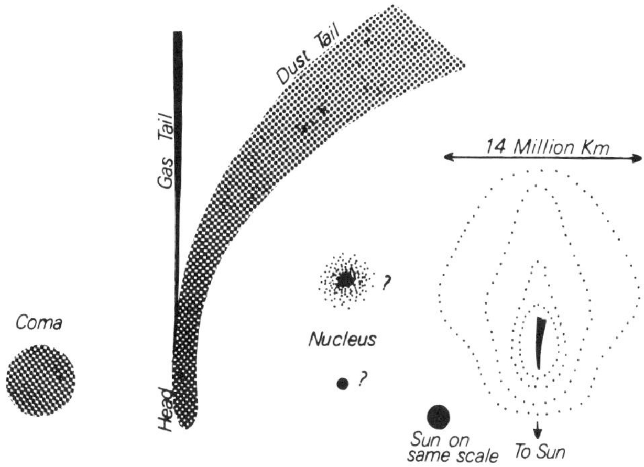

111 Structural components of a comet. Figure (*right*) shows the limits of the hydrogen gas envelope surrounding Bennett's comet (1970 II) as scanned by the Orbiting Geophysical Observatory-5. The different contour boundaries represent the varying densities of tenuous gas surrounding the normally visible parts of the comet (which is shown to scale).

112 Comet Bennett (1970 II) in a reverse print. (*see* also Fig 115).

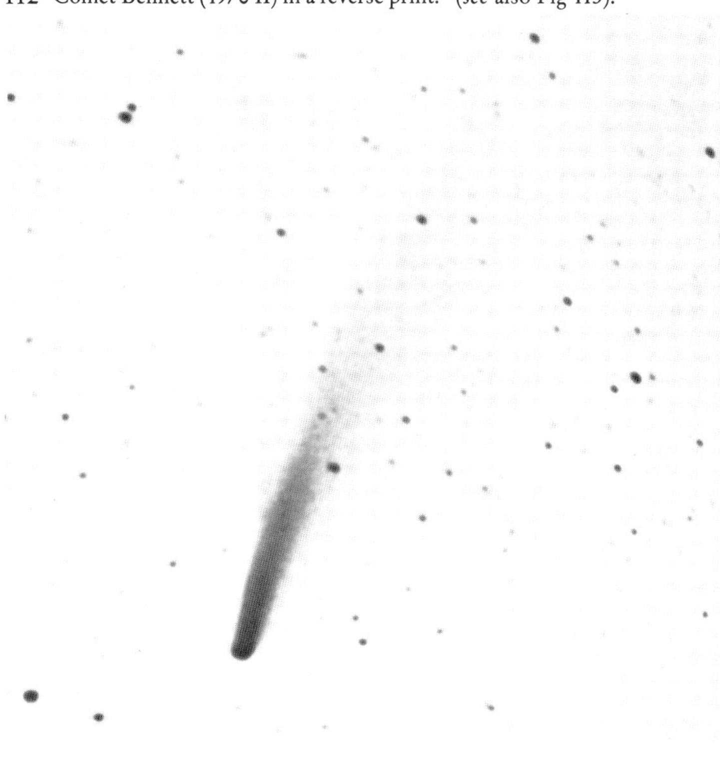

In William Herschel's time, astronomers of the eighteenth century often posed the question: are comets habitable worlds? Herschel himself, and others like him, believed they were. Later nineteenth-century astronomers had laughed this idea out of court, for now they knew better. Yet in the 1970s the subject would be aired again, although in a somewhat different guise when Fred Hoyle and his colleague Chandra Wickramasinghe re-posed the enquiry about life's origins in the form: do comets, in fact, provide travelling habitats for cosmic bacteria?

After the excitements of the year 1910 with the advent of two bright comets, in the years following, interest in comets languished, for astronomers were now preoccupied exploring the far depths of the universe and tended to neglect those objects nearer to home. This lack of interest was also brought about by a dearth of really bright comets. During the remaining first half of the twentieth century only a few bright comets were visible in favourable circumstances in the northern hemisphere, whereas during the nineteenth century there had been several runs of brilliant comets which attracted widespread attention.

There was a new interest in comets in the early 1950s, shortly after the discovery of the solar wind. This is the same emission which interacts with the Earth's magnetic field and produces brilliant displays of aurora in the upper atmosphere; it is also the same emission that can endanger the lives of astronauts in orbit round the Earth.

It was this particular emission emerging from the Sun which had greatly puzzled earlier astronomers. Now it was possible to understand the high velocity of ejected material in comet tails, for the bombarding solar wind itself travelled out from the Sun at speeds ranging from 1,000 to 2,000 km per second, plus a slower faction travelling at about 500 km per second.

Because it was found that comets were ultra-sensitive to this newly discovered corpuscular radiation, a new generation of astrophysicists began to use them as space barometers by which the solar wind itself might be studied.

Following the recognition of the role of the solar wind on comets, comet tails were classified into two main types: tails consisting of ionized gases (or plasmas), and tails consisting of a variable mixture of dust and quasi-dust particles resulting from the more simple radiation pressure of sunlight.

Both types of tails are easily recognizable in photographs. The ionized gas tail is generally straight and narrow (Fig 111) while the dust tail is strongly curved like a scimitar and tends to lag behind the line drawn between the comet and the Sun (the radius vector). The famous Donati comet of 1858 showed a magnificent tail-plume of what we now recognize to be "dusty" material, but it also shows two narrower, almost straight, tails of ionized gases (Fig 89). Thus the behaviour and origins of comet tails were by now fairly well understood, but what about the rest of a comet? What lay inside the shrouded nebulous head from where the tails were emitted?

By this time it was generally accepted that the structure inside a comet's head could be likened to a discrete, solid mixture composed of ices and dust in the form of a loosely knit conglomerate of matter. This provided the actual core, or nucleus, of the comet—often seen as a star-like object at or near the centre of the head—from where in bright comets, jets, or envelopes, of expanding nebulous material emerge.

This basic idea of an icy-conglomerate nucleus for comets had first been mooted back in the nineteenth century, but the modern and more advanced concept of it, circa 1950, was the independent brain-child of the American Fred Whipple, doyen among the post-World-War-II cometary astronomers.

Whipple visualized a comet as a kind of rotating "dirty snowball" made up of a discrete "hard" nucleus at its centre and packaged with such basic ingredients as water (H_2O), ammonia (NH_3), methane (CH_4), possibly carbon dioxide (CO_2) and dicyanogen (C_2H_2), plus other exotics and "solid" dust particles. In texture the nuclear structure might be likened to that of a yeasty raisin bread.

Not everyone agreed with this idea. An alternative nineteenth-century model had envisaged the structure of a comet to be a cloud of separate, loose (meteor) particles travelling together in orbit and compacted at the centre, providing the illusionary impression of a solid pseudo-nucleus. Perhaps not surprisingly, this particular comet model is nowadays appositely called the "flying sandbank".

With advances in the design and size of optical telescopes since the nineteenth century, there was always the chance that the question of which comet-model was the correct one might be settled by direct observation. In 1927, when the Pons-Winnecke comet came within 5.6 million km of the Earth, the French astronomer F Baldet observed it carefully with the 60-cm Meudon refractor of the Paris Observatory and concluded that the comet's nucleus, if it existed at all, could not be larger than a few hundred metres in diameter. More modern work by Elizabeth Roemer, using the 102-cm astrometric reflector of the Washington Naval Observatory, a telescope of exceptional definition, set an *upper limit* and indicated that comet nuclei as monolithic bodies could not be *larger* than about 3 to 4 km for short-period comets and 7 to 8 km for long-period comets. Comet nuclei, however, on average are likely much smaller structures.

Neither could the techniques of spectroscopy settle the question of the fundamental chemical make-up of comets by earth-bound observation. While great advances had been made in studying the physical constituents of distant stars and galaxies, comets remained enigmatic bodies. The spectroscope proved to be by no means infallible as a research tool in studying comets, for some emissions in comet spectra are known to be simulated phenomena which in reality do not exist at all. Parent molecules of the matter contained in the head are hard to identify. This is because they emit radiation in the infra-red and ultra-violet regions of the spectrum not available for study except outside the Earth's atmosphere. Because of the very low gas densities encountered in cometary atmospheres, molecules exist there in unfamiliar terrestrial forms—therefore posing a conundrum to those trying to identify them.

Apart from the puzzling chemical make-up of comets, another important unanswered question related to their origins. During the nineteenth century the belief that comets were captured by the Sun from space was much favoured; however, others preferred ideas for an origin within the solar system and that comets were formed of material ejected from the interiors of Jupiter and Saturn and/or from the interior of the Sun.

By the 1970s modern opinion swung to a more simple idea. Comets, it suggested, are partially condensed masses of primitive material—leftovers dating back to the time the planets were formed. Comets, therefore, today then represent the fossilized lighter factions of the original solar nebula and chemically are much like the material that went to form the outer planets Uranus and Neptune. Thus the large outer planets may be very near relatives to comets.

The Dutch astronomer J H Oort and his later disciples developed a theory that most of this remnant fossil material—known as "Oort's Cloud"—lies as a vast ring of orbiting comets at the very distant boundaries of the solar system. According to Oort's understanding we see new comets when nearby stars or massive planets like Uranus and Neptune perturb these distant comets and cause them to move inwards to make a rendezvous with the Sun. Thus those comets moving in the inner solar system are continually being added to from a vast reservoir of distant material.

In the mid-1960s radio astronomers had directed their large telescopes towards several comets in attempts to probe their physical make-up, but pioneer observations of the Ikeya-Seki Sungrazer comet of 1965 to try to identify microwave radiations from the ground proved unsuccessful. Then an entirely new tool came to hand in the form of the Orbiting Astronomical Observatory 2 (OAO2), launched on 7 December 1968...

The Earth's atmosphere acts as a shield, or filter, to certain radiations trying to enter it, and it had long been recognized by those making spectroscopic observations that fundamental knowledge about the physical processes taking place inside a comet's head would unlikely ever be forthcoming from earthbound observatories. If progress was to be made, the only solution was to place an observatory or laboratory outside the atmosphere. Until the era of artificial satellites this remained only the pipe-dream of astronomers. By the late sixties the old dream had become reality.

The first comet to come under scrutiny was the earlier of the two bright comets seen in 1969. This was discovered independently by a number of Japanese amateurs and later resolved itself

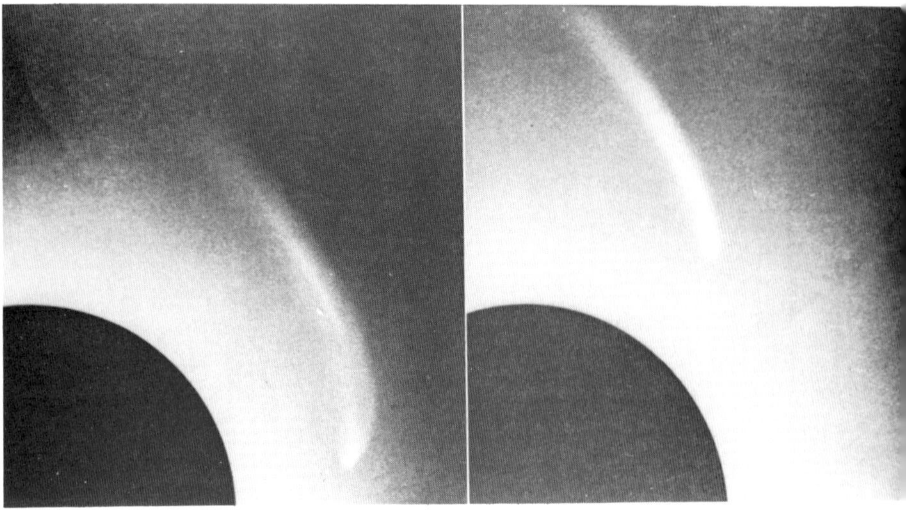

a)

b)

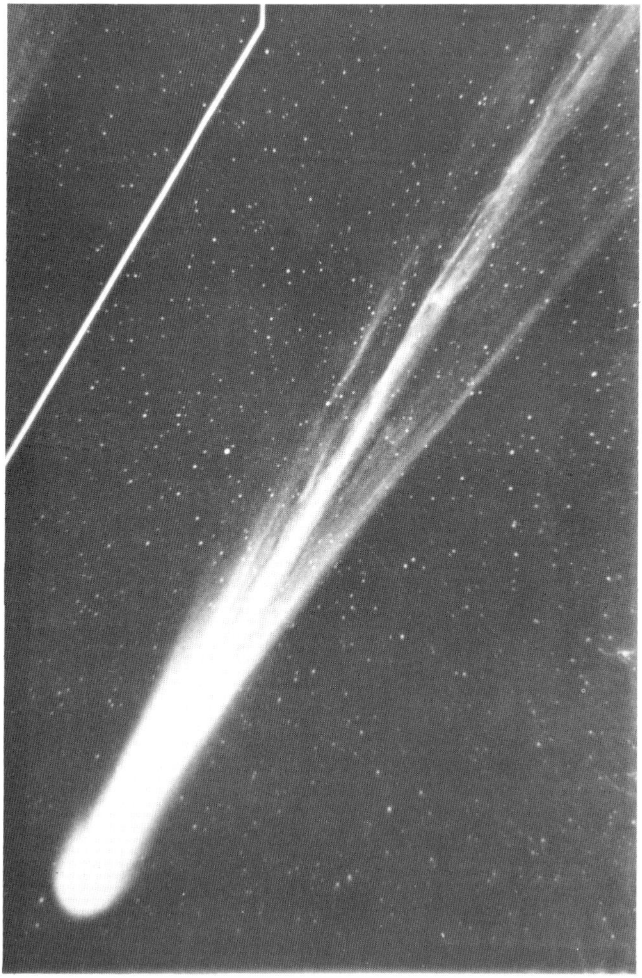

113 Comet Tago-Sato-Kosaka 1969 IX which, when viewed from space, provided the first evidence about the vast clouds of tenuous hydrogen gas surrounding comets (quite invisible in this photograph taken from a ground-level observatory).

114 The spectacular Ikeya-Seki comet 1965 VIII. This was one of the most remarkable Sungrazers ever seen and was plainly visible to the naked eye as it rounded the Sun on the afternoon on 21 October 1965. (*b*) When the comet emerged in the morning sky in early November, it provided some of the best views of a comet since the nineteenth century.

115 A spectacular display by Bennett's comet on 4 April 1970 showing disruptions in the gas tail (type 1).

in the catalogues under the title of comet Tago-Sato-Kosaka 1969 IX. The orbit of this comet proved to be very eccentric, and its period of revolution round the Sun is measured in some hundreds of thousands of years. Such an object, after visiting the Sun, would wander off into deep space, well beyond the farthest borders of the solar system, almost half-way towards the nearest stars before turning again for a rendezvous with the Sun. However, the most interesting fact about this comet, when observed from outside the Earth's atmosphere aboard OAO2, was the discovery of a vast cloud of tenuous hydrogen gas surrounding it, detected by the spacecraft's ultra-violet telescope. Comets were now, indeed, proved to be mostly gas bags and even more flimsy than thought in the beginning of the century.

The period from the middle 1960s to the middle 1970s had been a truly vintage decade for bright comets. Comet Kohoutek, in particular, discovered in March 1973, was intensively studied. As it approached the Sun, it became subject to wild speculations that it might become the brightest comet ever seen. While it did not live up to these expectations, it proved to be of great scientific interest. It was scanned from the ground and in the air in the ultra-violet, infra-red, radio and visual wavebands. The infra-red observations showed the tail had a dust content less than Bennett's comet—perhaps only one-sixteenth as much—and indicated that each comet is a unique object.

Emissions due to carbon (C_2), cyanogen (CN) , methylidyne (CH), the hydroxyle radical (OH) and the amino group (NH_2), all previously seen in emissions from other comets, were confirmed.

The presence of at least two organic molecules, methyl-cyanide (CH_3CN) and hydrogen cyanide (HCN) were inferred from radio-astronomical observations, plus several other organic molecules which had not been specifically identified. Fred Hoyle and his colleague Chandra Wickramasinghe were to make much of these observations when shortly after this they announced their startling theories about the role of comets in the distribution of life in space...

CHAPTER 18

Pandemics and Space Probes

Shortly after World War II, Halley's comet was moving at its farthest from the Sun, out beyond Neptune's orbit, and coasting at minimum speed. During the early 1950s it had slowly begun its long curve inwards so that by the mid-1970s it had reached the distance of Uranus and was picking up speed, accelerating sunwards to keep its next perihelion appointment, timed for early February 1986...

Since the comet had made its last appearance in 1909-11, research methods had changed out of all recognition. Scientists from many disciplines were now engaged in studying the various member bodies of the solar system as never before, and space probes and artificial satellites were sending back information totally inaccessible to previous generations.

While plans at this time were being quietly laid, a decade ahead, to mount the biggest comet watch ever for the mid-1980s, using space and earth-bound observatories, the subject of comets suddenly caused a furore in both the scientific and popular press and this had echoes far and wide. It stemmed from the ideas of two cosmologists, Fred Hoyle and his equally eminent colleague Chandra Wickramasinghe, who, as a result of their long-term investigation of interstellar grains, had come up with some sensational new theories. Foremost among their claims was the idea that comets were key bodies to a proper understanding of how life reached Earth in the distant past. Then to add fuel to their contention they enquired provocatively: "... was life still doing so today ... life in the form of primitive bacteria? ... And could viruses even be derived from comets?"

Hoyle, one of the world's leading cosmologists, had a reputation for stirring up controversy, but these latest suppositions, backed by Wickramasinghe, were even more breath-taking than his mastermind ideas for Stonehenge as an ancient observatory fostered in the mid-1960s. If the ideas of comets as transporters of life were true, it would mean vindication of the long-denounced astrological notions that comets were carriers of plague and other contagious diseases and give verisimilitude to John Gadbury's claims about the comet of 1664... It would overturn established ideas of how life evolved on Earth via a self-generated broth and how long-term Darwinian selection worked.

Even Hoyle and Wickramasinghe recognized that their most open-minded readership would have strong initial reservations about swallowing such heretical ideas. "The ideas seemed preposterous," the authors wrote, "but in science one must steel oneself not to decide the correctness or otherwise of ideas according to subjective prejudice. In science, fact reigns supreme..."

So what were the facts?

The authors' researches began in the 1960s with their determination to solve, as cosmologists, the mystery of interstellar grains lying out in space at large cosmic distances. These grains had been recognized for some time, and their presence became known because they caused a fogging of more distant stars on the photographic plate.

It was several years later before Hoyle and Wickramasinghe, having in turn considered and then rejected a whole inventory of possible substances, finally concluded the grains were organic in origin. Their supposition, they believed, was backed up by the remarkable discoveries of radio telescopes which had detected emission sources from deep space for several organic molecules. The advent of comet Kohoutek in 1973-4 seems also to have channelled their thinking. This comet had contained a diverse range of material, including some organic molecules (*see* p223). All in all it was a very impressive package and enough to indicate to more fertile minds that comets, instead of being insignificant members of the universe as some supposed, might play a key role in the distribution of exotic material throughout space...

In 1977, Hoyle and Wickramasinghe crossed their personal Rubicon. It was now they stepped across the boundary from orthodox science into the realms of the heretical non-orthodox.

They claimed that the puzzling grains out in space not only contained complex organic molecules like those detected by radio telescopes *but were in fact bacteria-like organisms.* Their grain size, they claimed, was just about right to match the spectral characteristics of bacterial cells studied on Earth.

It was from this general concept that Hoyle and Wickramasinghe developed their life-from-space ideas through a series of semi-popular books. Their initial premise was to show that contrary to established thinking, which had life on Earth evolving from primitive self-generating biological broth, *it arrived ready-made from space.*

While Hoyle's and Wickramasinghe's ideas about the nature of interstellar grains were provocatively contentious to those astrophysicists engaged in similar work, they were by no means the first to advocate life-from-space ideas. In the past, meteorite falls had often been cited as possible examples of how life might have seeded the Earth. Several meteorites had been suspected by some chemists of containing organic-like substances, and in 1871 Lord Kelvin, in his address to the British Association, reminded his audience that there was a real possibility that life could have arrived on Earth this way. Some of those present, however, believed they were victim to an outrageous presidential joke and treated it as such. But there is little doubt Kelvin was in deadly earnest.

Another who was in deadly earnest was the Swedish chemist and Nobel Laureate Svante Arrhenius, who in 1907 published a book *Worlds in the Making* in which he proposed his panspermia hypothesis, suggesting that the simple forms of life had drifted from world to world propelled by radiation pressure of sunlight through the medium of interstellar space. It was just before this time that (simple) radiation pressure from the Sun was identified as probably very significant in producing cometary (dust) tails, and the concept was seized upon as an ideal motive force operative in "empty" space. Other writers, in particular the German

117 Comet Kohoutek photographed on 13 January 1974. Although, from the public's point of view, this comet proved a disappointing spectacle, it stimulated Hoyle's and Wickramasinghe's ideas about the role of comets in distributing life in space.

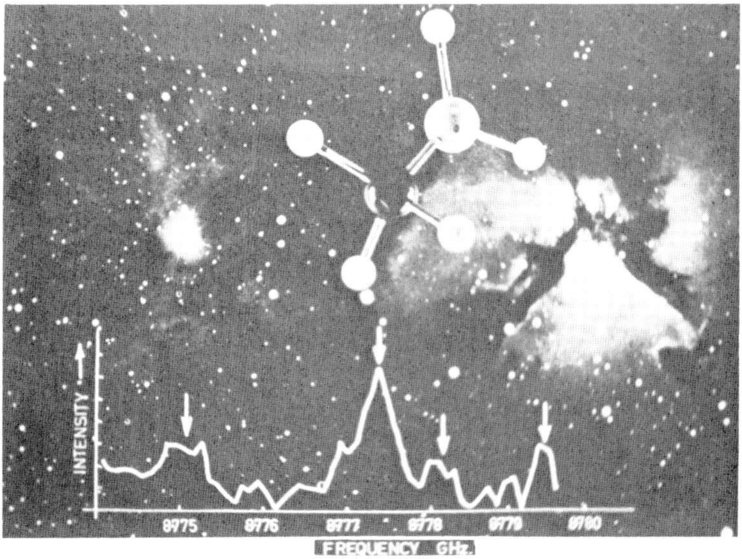

116 The signal and model of the organic molecule methylamine superimposed on a photograph of the Trifid Nebula (near the centre of the Milky Way) in which the source was detected by the Parkes radio telescope in Australia. Such molecules may also be significant in comets.

scientist H E Richter, suggested that life from Earth might itself be transferred to other worlds by meteorites brushing with and then bouncing off our atmosphere during which time they picked up living cells.

The authors of all early ideas were handicapped by lack of knowledge of conditions existing in space. It was wondered how living cells wandering freely from star to star, propelled only by radiation pressure, could survive the hostile environment on long journeys. It was pointed out that ultra-violet radiation would be lethal to any extraterrestrial spores exposed in space. When the even more deadly cosmic rays were identified, unknown in Arrhenius's time, these theories of life transference were damned and consigned to the scientific scrap-heap. There they remained until resurrected by Hoyle and Wickramasinghe circa the late 1970s.

Hoyle and Wickramasinghe carefully avoided the older meteorite hypothesis. Although in the 1960s and 70s several other meteorites had fallen containing substances resembling organic compounds, meteorites *per se* had great limitations as carriers of cosmic spores from far space. Instead they opted for comets, no doubt having in mind the rich onboard package identified in comet Kohoutek. Comets had the advantage they moved in highly eccentric orbits which led them at aphelion to pass beyond the normal outer regions of the solar system and then swing back towards the Sun. When they were close to the Sun and the inner planets, they were subject to strong radiation pressure as well as bombardment from corpuscular particles forming the solar wind. Might not the dusty tails of comets be the actual spreaders of cosmic spores and bacteria that would then be carried into the Earth's atmosphere as micro-meteorites when the Earth sweeps through the tail of a comet or intersects its path?

Hoyle and Wickramasinghe, as cosmologists, had never previously been much interested in comets. Like others in the same field they had considered comets as playing but a minor role in the shaping of the cosmos. Now, however, things were different, comets could be requisitioned as the ideal vehicles for transporting those all-important interstellar grains to the inner regions of the solar system on a periodic basis. Their arguments took the

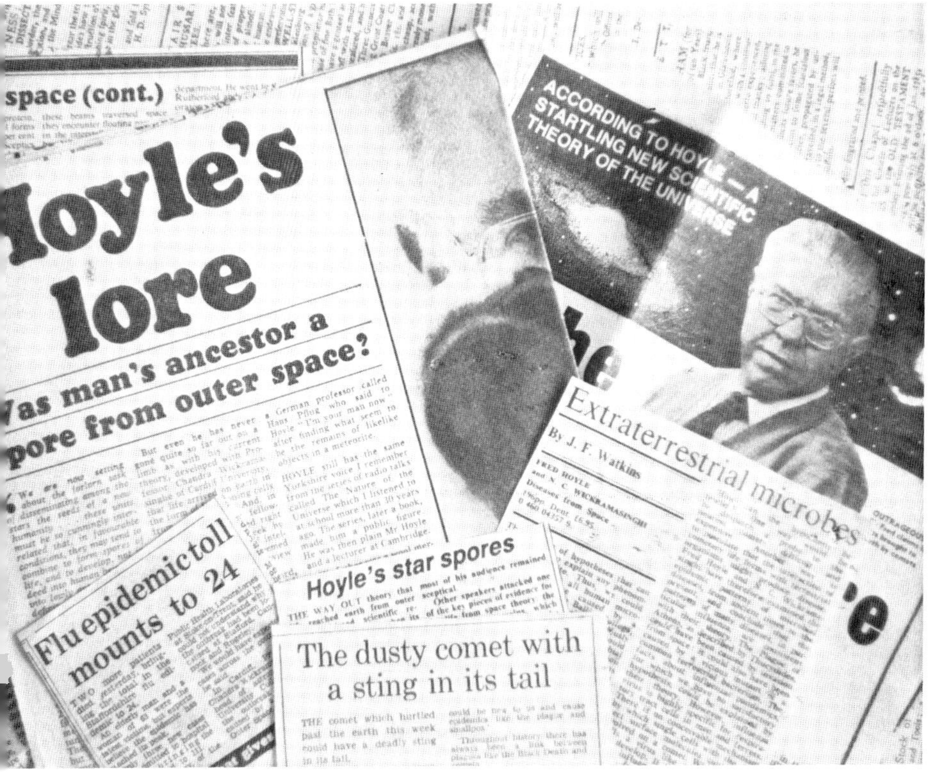

118 When Hoyle's and Wickramasinghe's theories of life from space were announced, they created a considerable stir in the popular press.

line that when the solar system was created 4.6 eons ago, hundreds of millions of comets were formed as ice-balls and remained circling the outer regions. Subsequently, on the surface of these ice-balls (comets) were deposited interstellar grains which grew as mantles—perhaps up to one kilometre thick. Collisions between comets as they were jostled about would release energy; any melt-water trapped near the ice-core boundary would then provide liquid nurseries for cosmic spores and bacteria that could remain preserved for millions of years.

In a later development of their idea, Hoyle and Wickramasinghe saw comets more as *carriers* rather than as original reservoirs of genetic material. Either way, these onboard packages remain preserved long enough to survive the day when their host in outer space is perturbed by a passing star (the Oort hypothesis) and starts to head inwards for a rendezvous with the

Sun and Earth. Thus comets transformed into short periods, like Halley's comet, would periodically dissipate their organic cargoes under the erosive influence of sunlight. To explain germ-spread and resultant diseases borne to Earth, Hoyle and Wickramasinghe cited, in the case of Halley's comet, that pandemics could be expected at intervals varying from 74 to 79 years to match variations in its orbital period.

The question raised is whether any of these ideas do have verisimilitude. It seems ironic that Halley and his peers at the Royal Society were very sceptical about Gadbury's claims relating plagues to comets, yet 300 years on Sir Fred, as one of the Society's most illustrious scientists, was vindicating the very same heresies! However, Hoyle's prime target was not simply linking diseases to comets but a frontal attack on the tenets of Darwinian evolution. Not only did Hoyle and his colleague see life on Earth as a product of seeding from space from the very beginning, they also maintained it continued to reach the Earth all through time—battering at the door of our planet into the immediate present and providing the mechanism that determined evolution and the subsequent diversification of incumbent species.

The fact that Hoyle is a superb mathematician, able to manipulate numbers at the drop of a hat to support his ideas, makes him a formidable target to attack for non-numerate biologists who by a very large majority totally oppose his evolution thesis. His often statistical approach to evolution, citing as he does that Darwin's theory is mathematically impossible, makes it difficult for critics to answer him in relevant numerate counter-arguments. While Hoyle, in particular, has a succinct way of marshalling a battery of cool, logical argument, he is sometimes careless with his facts, often misquoting and leaving himself vulnerable.

Hoyle and Wickramasinghe believe that biologists who opt for life's origins on Earth take a narrow, pre-Copernican view of the matter. They point out that space offers infinitely more opportunities as a synthesizing laboratory for life than did a primitive Earth.

Nevertheless, many distinguished scientists refrain from

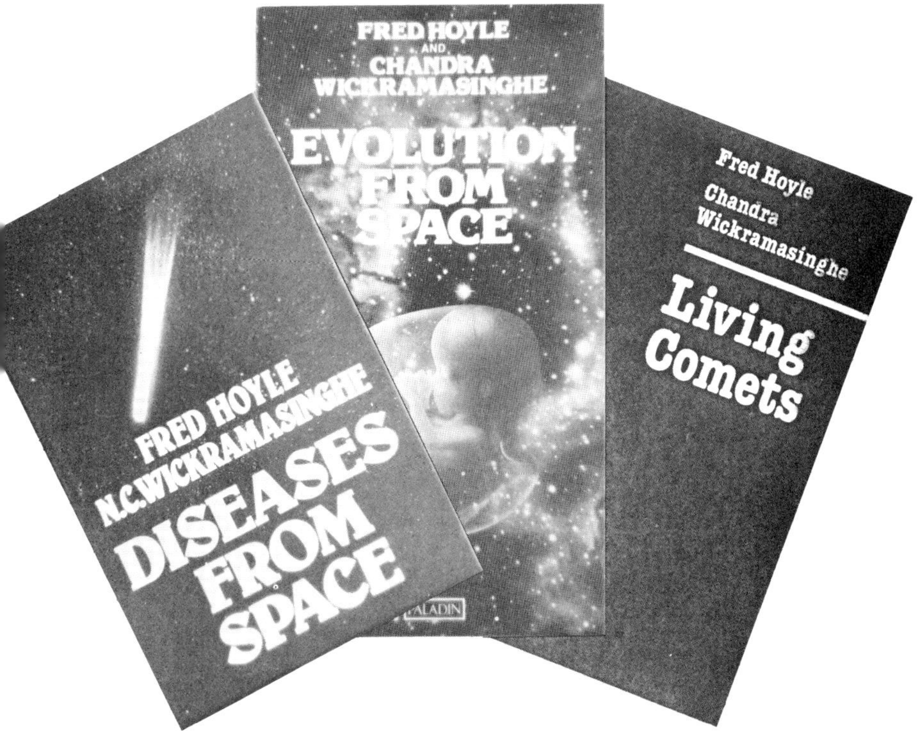

119 Sir Fred Hoyle and Chandra Wickramasinghe announced their startling ideas about life and diseases from space via several controversial books, three of which are shown here.

offering any kind of criticism and prefer the wall-of-silence approach to register their disdain of Hoyle's and Wickramasinghe's entire thesis. In their view it is no more scientific than Velikovsky's book *Worlds in Collision* which created a furore in the early 1950s. They believe that offering criticism is time-wasting and nugatory. They emphasize the two authors repeatedly misrepresent evidence from space as well as Darwin's ideas, and in respect to these they are totally ill-equipped by their training as astrophysicists to dabble in areas beyond their own discipline.

Until unambiguous evidence is forthcoming via space probes, Hoyle's and Wickramasinghe's thesis for the role of comets remains unproven. Among their claims, it is the origins of pathogens which meet with a very formidable opposition. Yet biologists and virologists have themselves puzzled over these origins and modes of distribution. Nevertheless, it seems unthinkable that bubonic plague, influenza, Legionnaires' Disease ... and perhaps the most dreaded of all, the virulent AIDS, are products derived from cometary encounters.

But in spite of the battery of criticisms against, the search for life in comets offered exciting possibilities for fly-bys to Halley's and other comets. Whether the question can be answered in the immediate future when biological samplers are landed on a comet nucleus is uncertain, but techniques had already progressed to the stage where such missions to comets were already almost a practical realization (*see* p 267).

Since the last appearance of Halley's comet, advancing technology had brought a revolution in the design of computers. The mechancial hand-machines that Cowell and Crommelin and their three helpers had used to solve the complex equations of motion for the comet were now museum pieces. Tasks which earlier in the century had taken the British team weeks of labour to calculate could be programmed and then printed out on the new-generation electronic machines in fractions of a minute. While the tedium was now removed from the exercise, the new super-fast machines were still only artefacts; they could only perform within the limits of accuracy of the observational data fed into them. With the onset of the 1986 apparition of Halley's comet there still remained the problem surrounding the magnitude of the non-gravitational forces at work on it that had sorely puzzled the British team and had led to a disparity of three days between prediction and actual perihelion date.

The non-gravitational forces operative in comets, but not predictable by Kepler's laws, were by now better understood. They were believed to be caused by an outgassing in the comet's icy nucleus producing a rocket-effect leading to changes in orbital speed. Indeed, the very presence of non-gravitational forces operative in comets was taken to confirm *a priori* Whipple's idea of a discrete icy nucleus, for it was difficult to see how a rocket-effect might arise in a loose agglomeration of particles as proposed in the alternative sandbank model for comets.

During the 1960s the first plans were drawn up by various

agencies for future fly-by missions to suitable periodic comets. In the late 60s and early 70s, space vehicles in orbit looking at comets outside the atmosphere discovered the vast envelopes of hydrogen gas surrounding them—invisible from earth-bound observatories (p 219). It was clear by this time that a close fly-by mission or an actual rendezvous with the inner nuclear regions was the only hope of ever determining the true physical make-up of cometary bodies and their possible close relatives, the Earth-crossing Apollo asteroids considered by some to be old comets now defunct of their nebulous envelopes.

The prime target for the future missions was Halley's comet in 1985-6. Apart from the fact it was the most famous of all comets, it was thought to be large in comparison with most periodic comets. Other periodic (P) comets short-listed were P/Encke which was favourably placed in 1974 and again in 1980; P/d'Arrest in 1976; P/Kopff in 1983 (later rescheduled for a definitive fly-by in 1994); and P/Arend-Rigaux in 1984.

Clearly a practice-run at *any* comet before 1985-6 if this could be arranged would help iron out some wrinkles for the later probe to Halley's. In the interim years, however, constraints by NASA on budgets decided them to concentrate their efforts instead on planetary explorations using Mariner, Pioneer and Voyager probes. The Soviet Union likewise, long interested in comet probes, also limited its activities to planetary probes.

It was soon clear to the comet-mission planners that compared with planetary missions those to comets faced several difficulties. A major problem was pinpointing a comet's path with sufficient precision to make an intercept. In most cases the non-gravitational forces were small yet significant enough if not allowed for; and if a comet travelled anywhere near Jupiter before its next rendezvous with the Sun, the perturbing effects of the giant planet could make the calculations for its intercept, without later corrections, a hit-and-miss exercise.

Several comets were short-listed for space missions, but Halley's comet remained the primary target.

Earlier in the century, at its previous apparition, Halley's comet had been spotted 249 days before perihelion; then after perihelion, in mid-April 1910, it was followed at Helwan for 396 days, making a total of 645 observation days. This, however, was surprisingly 5 days less than in the 1835-36 apparition.

If the comet could be resighted about 400 days before perihelion, it would provide time for a series of pre-launch checks so that corrections could be incorporated into the ephemeris to make targeting more accurate. Thus provisionally a very conservative expectation date for recovery was set for January 1985. By the second half of the 1970s, observers were a good deal more ambitious than this figure suggested...

Since the turn of the century, telescopes and their accessories had made enormous technological advances in design in common with computers. Even the older telescopes were now equipped with electronic gadgetry able to record objects way beyond the faintest vision of the observers who had used them just a few decades earlier.

Yet, in spite of the extreme sensitivity of modern instruments, it still came as a surprise when observers with access to two of the largest telescopes in the United States announced their searches for the comet had begun as early as November 1977—eight years before perihelion date. At this time Halley's comet was estimated to be dimmer than 26^{m}. It was abundantly clear that the competitive spirit of yesteryear lived on and there were those anxious to add their names to Farmer Palitzsch, Father Dumouchel and Max Wolf.

For the 1985-6 apparition several mathematicians made independent computations to retrace the comet through previous apparitions to try to determine the exact time for the new perihelion due in early February 1986.

One of the major studies originated from the NASA Jet Propulsion Laboratory at Pasadena and was published by Donald K Yeomans under the title *An Observer's Guide*. This included an ephemeris for the comet for 1981-87 to assist those observatories and astronomers participating in the "International Halley Watch" programme organized under the auspices of the International Astronomical Union. Yeomans utilized a total of 885 past observations beginning with one of Kepler's made on 28 September 1607 and ending with a photographic position for 24 May 1911. In his comprehensive analysis of the comet's path, spanning the four centuries of motion since Kepler's time, Yeomans also made adjustments for the factor of non-gravitational forces operative on the comet.

An even more in-depth study than the above was one made some time previous by Tao Kiang, who retraced the path of the comet through earlier Chinese observations back to 240 BC. Kiang's interest in the comet had been triggered in 1969 after becoming intrigued by Ho Peng Yoke's comet catalogue, which provided a set of unique observations on ten specific dates for the return of Halley's comet in AD 837 and not available to Cowell and Crommelin earlier in the century. Assisted by a modern computer, Kiang repeated Cowell's and Crommelin's pre-electronic-age calculations. He adopted much the same methods but now with much better equipment than his predecessors incorporated in them a number of refinements that included perturbations by all the major planets in all three dimensions.

Kiang, because of his background, was able to resolve early oriental texts at first hand. By adopting a mixture of computative and graphical analysis, he traced the orbit back through each apparition. Much of his time was spent on a detailed re-examination of the original records rather than on second-hand Western translations which others had usually relied on. He was able to clear up several previous ambiguities respecting early Chinese observations and concluded that by the ninth century AD Chinese astronomers had been able to measure a comet's position to an accuracy of one half of a degree, a feat not accomplished in Europe until well after the Renaissance.

Kiang confirmed that the return of AD 837 had been a very noteworthy one. On 9 April that year the comet approached the Earth to within 0.04 AU and moved 60° across the sky in 24 hours. As a result of this very close approach (equal to ten times the Moon's distance), the Earth's gravitational pull greatly affected the comet's motion. Chinese observers reported its tail to be over "80 feet" long or, as Kiang translated it, equivalent to more than 100° in length, rivalling the tail length reported in 1910 but not then seen to best advantage. His researches also shed light on the length of the tail during its famous apparition in 1066. Chinese observations had mentioned the comet passing through various stars, and Kiang interpreted this as their method of describing *tail length* rather than the movement of the comet itself. Applying this idea to what the observers actually saw on 24 April 1066, it indicated the tail then stretched from horizon to horizon and hung northwards, passing directly through the north celestial pole. Next day the tail was still a magnificent sight as it swung round and spread itself across the constellations of Auriga, Gemini, Leo, Virgo, Libra and Scorpius.

A third mathematician to investigate the orbit of Halley's comet was Joseph L Brady whose work extended the apparitions back to 2647 BC. Some years earlier, Brady claimed anomalies in the motion of Halley's comet were due, in part, to the presence of a yet undiscovered massive planet lying beyond Pluto. Although he computed a rough ephemeris for the supposed path of this body, all searches made for it had so far proved negative.

In respect to his work on the comet, Brady drew particular attention to the problem that all previous efforts to link past apparitions relying only on Newtonian equations had failed each time. This, he pointed out, had resulted in at least a three- or four-day error in predicting the time of perihelion, and the discrepancies always had the same directional bias. By including in his own calculations what he referred to as a secular term (i.e. an allowance for non-gravitational forces) to traditional equations of motion, he claimed he and his colleagues had successfully linked back, making use of about 5,000 observations, the four apparitions of 1910, 1835, 1759 and 1682.

By continuing the integration programme further back in time, Brady also claimed to have fixed earlier apparitions with a numerical accuracy greater than that found in ancient observational records. The scope of this ambitious work included retracing 34 apparitions *previous* to that of 87 BC, the oldest date for Halley's comet identified by Cowell and Crommelin with certainty. Thus the comet's supposed history was taken back to 2646 BC.

In apparitions before 240 BC, which was Kiang's limit based on reliable Chinese observations, Brady admitted that any confirmation of a sighting of a comet with a computed apparition in the appropriate year is beset by the problem of recognizing in

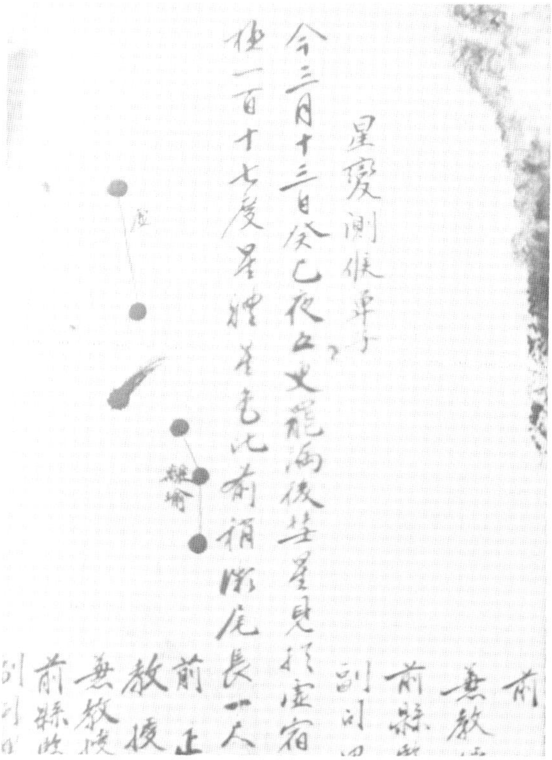

120 A Chinese manuscript recording the appearance of Halley's comet in 1759.

inscriptions what he refers to as "the cometary metaphor". Even this admission understates the difficulties. Scholars have long appreciated one may read into early Babylonian texts any manner of phenomena to suit a particular theory and they frequently, and hotly, dispute each others' claims what a particular inscription may actually refer to. In the later Babylonian astronomical diaries, where accounts are much clearer, *there is not a single reference to a comet to be found anywhere!** In older, more ambiguous texts, some of which are only hastily made copies of earlier (also copied) texts and contain errors and lacunas, it is difficult even to disentangle planetary names and the names of a few stars and constellations.

*But *see Endeavour* (VOL 10, No 2, p 95) for a contrary claim.

Yeomans and Kiang together had also performed much the same back-tracking exercise as Brady and speculatively traced out the comet's supposed apparitions over 3,000 years. Yeomans's approach was more cautious than Brady's, and he stopped his integration at what he claimed was a return in 1404 BC where Brady cites the year 1445 BC. This difference of over 41 years highlights the problem in retracing a comet, even with the help of modern computers, *without* sure observations for it.

Brady's less conventional methodology involved the use of what can only be termed a variable constant (a contradiction in terms), and this factor certainly led to differences in cited perihelion dates back in prehistory. Brady, perhaps with some justification, explained *some* differences in perihelion times from those of Kiang's (based on actual Chinese observations) were due to the fact that the ancient Chinese calendrical systems could not be easily correlated with the Western Julian and Gregorian calendars as some might suppose.

Back in prehistory, according to Brady's calculations, the comet sometimes returned to perihelion in a shorter period than the *observed* all-time record of 74.42 years between 1835 and 1910. For example, the return in 1646 BC supposedly took only 73.08 years (*see* note 28).

By January 1981, Halley's comet was moving between the orbits of Uranus and Saturn and gradually picking up speed, but it had still not been resighted. Its inward path lay among the very rich star fields of the Milky Way on the borders of the constellations of Canis Minor and Monoceros. As seen from Earth its daily shift among the background stars would be very slow. Its predicted brightness round this time was estimated to be still fainter than 25^{m}. While theoretically the comet was now just within range of the largest modern telescopes, equipped with image-intensifiers, the task of identifying it from among thousands of other faint anonymous stellar images at the limit of visibility was like searching for the proverbial needle.

By this time the comet-mission planners, who were still considering options for a probe to Halley's comet, were feverishly mounting their behind-the-scenes programmes. In the interim,

NASA, to the surprise of all, had declared it had no firm plans for a unique Halley mission. They were in fact reserving their options for a pre-emptive fly-by mission to the very favourably placed P/Giacobini-Zinner in 1985, using a probe named International Comet Explorer, which, it was estimated, would also be able to pass within 30 million km of Halley's comet in 1986.

Those who did have firm plans for Halley's comet as a prime target were the European Space Agency (ESA), the Japanese Institute of Space and Astronomical Science (ISAS) and the Soviet Union (*see* Table 3 and Chapter 20).

Table 3
Missions to Halley's Comet 1985-6

Mission	*Country*	*Launch date*	*Fly-by date*	*Fly-by distance*
Vega-1	USSR	15 December 1984	6 March 1986	10 000 km
Vega-2	USSR	21 December 1984	9 March 1986	10 000 km or closer
MS-T5	Japan	5 January 1985	8 March 1986	4-5 million km
Giotto	ESA	2 July 1985	13 March 1986	500 km
Planet-A	Japan	14 August 1985	8 March 1986	200 000 km

Although the search for the returning comet had begun in 1977, by 1981 there was still no sign of it. It would now be accelerating in towards Saturn's orbit, but so far the myriads of faint background stars had rendered it anonymous. Even with an object as important as Halley's comet, the time allocated for observing it was severely limited by the constraints of observatory schedules.

Telescopes known to be involved in the hunt included the giant Soviet 6-metre telescope, the 5.08-metre Hale telescope at Mt Palomar, a 4.5-metre telescope in Arizona, three 4-metre telescopes in Arizona, others of similar size in Australia and South America, a 3-metre at the Lick Observatory and the 2.7-metre telescope at the MacDonald Observatory. Never before had such a battery of large telescopes been brought to bear in efforts to resight a returning comet.

In December 1981, in the transparent mountain air at the Kitt Peak Observatory in Arizona, a concerted attempt was finally made to hunt down the elusive comet, using the 4-metre telescope coupled to a sensitive electronic image-intensifier. This arrangement was capable of picking up any faint object to a limit of $24^{\mathrm{m}}.3$. Nevertheless, even now the telescope and its advanced equipment failed to reveal any slow-moving stranger lurking among the star fields.

The months rolled by, 1981 turned into 1982; then in the autumn of 1982, two keen astronomers at Mount Palomar, G Edward Danielson and David C Jewitt, arranged to "borrow" from their colleagues a slot of observing time with the 5.08-metre Hale telescope already scheduled for other programmes of work.

Nowadays, in common with observatories elsewhere in the world, the night skies round Mt Palomar are plagued by sky fogging owing to street lighting in nearby towns. When, however, sky conditions are good, the great Hale telescope is still a superb performer and coupled with an image-intensifier is capable of detecting objects to the limit of 25^{m}.

The first early-morning session made available was on 16 October 1982 when the pair installed an electric charge-coupled device (an intensifier) at the prime focus of the 5.08-metre reflector. At this time it was calculated the comet was at a distance of 1.6 thousand million kilometres from the Sun, and according to prediction should be slightly brighter than in December 1981. Over an interval of $1\frac{1}{2}$ hours the two observers made five eight-minute exposures in yellow light over the small region of the sky where prediction indicated the comet should be. Then, when twilight was rapidly coming on and their observing time fast running out, they hurried to obtain two further exposures but this time in red light.

By the end of their early-morning session, Danielson and Jewitt were holding their breath. A quick check had revealed that on each exposure they had made there was indeed a tiny slow-moving image which approximated to the predicted ephemeris. The images, however, were very faint and estimated to be about $24^{\mathrm{m}}.2$, just above the threshold limit of the telescope.

After alerting Brian Marsden, incumbent Director at the IAU clearing-house for astronomical discoveries located at Cambridge, Massachusetts, the problem now was deciding whether the images were those of the comet or alternatively those of a previously undiscovered faint asteroid. This latter alternative was a real possibility and could perhaps explain the shift in motion of

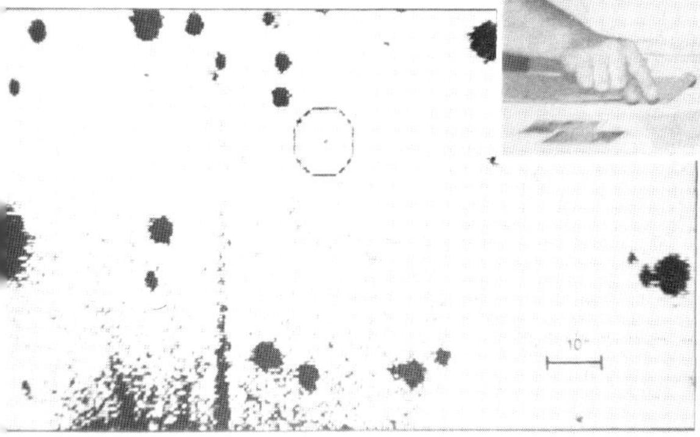

121 (*above*) The first glimmer of Halley's comet to register itself after an absence of 71 years. On October 1982, two astronomers, G E Danielson and D C Jewitt (*right*) using the 5.08-metre Mt Palomar reflector (*illustrated*) succeeded in recording the tiny image of the returning comet (*circled*) three years and four months before it reached perihelion on 9 February 1986. At the time of recovery the comet still lay outside Saturn's orbit and at this distance was about 3,000 times fainter than when first recovered in 1909.

about $3\frac{1}{2}$ seconds of arc over which the object had travelled during the course of their observations. However, there was a sure clue about the motion of Halley's comet: it moves in a retrograde orbit, the opposite way round to that of the major planets and the asteroids. In the part of the sky observed by Jewitt and Danielson, any asteroid, Marsden decided, would appear to shift *eastwards*; but the object found was shifting towards the *south-west*. In Marsden's opinion there was little doubt that what the two Palomar observers had picked up was the comet.

In 1758, Palitzsch had resighted Halley's comet less than three months ahead of its perihelion; in 1835, Father Dumouchel's observations were just over three months ahead; while in 1909, Max Wolf photographed it seven-and-a-half months ahead. This time round the two Palomar astronomers had broken all records by finding it three years and four months ahead.

But sure confirmation was required. Attempts were made three mornings later with the same telescope. As expected, no images were found in the places previously occupied by the moving object on 16 October, and this absence added confirmation to the reality of the earlier observations. Unfortunately, by now the comet was too close to one of the brighter field stars whose dazzling image overpowered any feeble light emitted by the comet. In due course, however, confirmation was made, and when its position was known exactly, other telescopes of much smaller aperture were able to photograph it and then keep track of it.

Check calculations using the observed positions showed that those who had taken into account non-gravitational forces had this time almost got it right. The comet would arrive at perihelion 9.3 UT February 1986. Yeomans had predicted a time of 9.66 UT February 1986 while Brady had obtained an even closer result of 9.39 UT February 1986. An earlier prediction of 5.37 UT February 1986, calculated by a mathematician several decades earlier, but using only Newtonian theory without a correction factor, seems to confirm that non-gravitational forces, at present amounting to 3-4 days per revolution, are still operative on the comet and affecting it by the amount Cowell and Crommelin themselves suspected back in 1909. (*see* p 322)

CHAPTER 19

In the Days of the Comet

"A single comet begets a thousand fools."

—SENECA

"There's money in mania."

—Overheard in Wall Street

After the flurry of interest caused by the recovery of Halley's comet in October 1982, comet watchers prepared themselves for a long three-year wait until it drew closer. The comet, it was predicted, would not be sufficiently bright to gain a wider viewing audience before the autumn of 1985, and it would not become even a faint naked-eye object until December that year. Nevertheless, during its long unspectacular run-in, the comet soon began to attract the almost daily attentions of the popular media.

Since the comet had departed in 1911, it was not only computers and telescopes which had undergone radical developments. This time round, Halley's comet was returning to a world dominated by television and radio—a world plagued by commercial hype and brash advertising. While in 1910 the comet had been subject to commercial exploitation (Fig 103), this was insignificant in comparison with what now awaited it. Indeed, *Punch*, in 1910 (Fig 102) had hinted that "Halley" personified the fast diminishing world of the amateur and with it the implications of an increase in bad taste. By 1985-6 these old fears were fully realized.

The new-style 1980s' hucksters quickly identified Halley's comet as a heaven-sent money-making opportunity—a returning celestial spectacle that could be used to exploit an ever gullible public now with a potential spending-power quite unknown in previous ages...

There was precedent to all the razzmatazz which was soon to

envelop Halley's comet. During the seventies, Kohoutek's ill-fated comet had for a time attracted wide-spread publicity. Because of its unusual brightness at the time of its discovery in early 1973, a forecast was made (somewhat rashly) that this comet, by the time it reached perihelion at the end of the year, would become the brightest comet ever seen.

It was now the popular media, fed with these highly optimistic predictions by attention-seeking astronomers who should (on past experience) have heeded caution, prepared the public for a potentially unique event that threatened to rival the legendary comet seen during Mithridates's reign (note 27).

Soon condoning this wild optimism were publicists working with NASA, who recognized it as an opportunity to justify the large public expense then being incurred (and strongly criticised) in the Skylab programme at that time running at $4 million a week. However, in spite of this political motive, the approaching comet was indeed a very suitable target for study outside the Earth's atmosphere by a satellite.

As the date for the comet's perihelion drew closer, the band-wagon of optimism in the popular press, on television and radio gathered momentum and notwithstanding some increasing doubts expressed by more rational astronomers, it became unstoppable. By now the event was promoted to the status of a Barnum and Bailey production. Fancy-goods and T-shirt manufacturers had almost overnight gone into round-the-clock schedules in preparation for the forthcoming celestial rip-off. One British company went so far as to register itself as Comet Kohoutek Ltd in attempts to corner the market brand name.

The cranks, as always, had a field-day and printed a spate of doom-watch pamphlets. These were echoes of all those which had appeared since the advent of the Plague and Fire comets of 1664 and 1665 and confidently predicted the end of the world. For the first time in history (and perhaps, unfortunately, the last) comet books, well-informed or otherwise, were selling in New York as fast as the latest Pulitzer Prize winner. Airlines and travel agencies, also keen to get in on the act, promoted charter-flights for the comet buffs who did not want to miss out because of last-minute bad weather or cloud-outs. Others were enticed aloft on charter-flights with tongue-in-cheek promises that the comet would appear much brighter to spectators from 30,000 feet plus than from ground level. This dubious argument, which glossed over the impracticalities of observing through the small thick windows of modern passenger jets, was dusted off again by airline companies promoting flights sponsored to conveniently fill empty aircraft when Halley's comet came round in 1985-6 (*see* below).

Meantime, Lubos Kohoutek, the Czech-born discoverer of the comet and himself a very modest astronomer and innocent party to all the fuss, was (as a lecturer) persuaded aboard the luxury cruise-line *Q E 2* specially commissioned to sail out of New York for the big occasion. His potential audience was a nouveau-riche clientele who, fed with blandishments they would be truly privileged comet watchers, took passage on the shipping company's recommendation that a view from the darkened skies of the North Atlantic would be a "once-in-a-lifetime" sight. During 1985-6, this "once-in-a-lifetime" cliche was again to be exploited to the point of banality by authors and hucksters alike.

On schedule, at the end of December 1973, the Skylab crew observed Kohoutek's comet from space just after its perihelion passage. Although not a tenth as bright as some had rashly forecast, it did for a short time become a relatively prominent object from space. However, at this time it was badly placed for public viewing both at ground level and from the air.

The media soon suspected something was wrong, and it was now they accused astronomers of letting them down.

By the time the wider public had chance to view the comet, as it slowly emerged into the evening sky, it had dimmed to the general appearance of a fuzzy fourth-magnitude star, not even properly visible to the naked eye. Out in the storm-tossed North Atlantic, the special comet-watch cruise of the *Q E 2* was persistently dogged by bad weather and cloud, and for much of the time her passengers, including the discoverer himself and fellow lecturer Isaac Azimov, were below decks hors de combat with sea-sickness.

From the point of view of the man in the street long prepared to view a comet spectacular, Kohoutek proved a very damp squib. As a result the public for a time were now very wary and even downright distrusting of *all* astronomers' predictions. Consequently when comet West graced the morning skies in early March 1976, few could be tempted to desert their warm beds. Thus the doubting Thomases all missed out on a magnificent spectacle that even outrivalled the display put on by comet Bennett a few years earlier.

122 Comet West gracing the dawn sky on 4 March 1976.

In contrast to the earlier, quite outlandish, forecasts for Kohoutek's comet, Halley's comet on its return in 1985-6 had been more soberly predicted as providing a potentially poor public showing. This was especially emphasized for those in the northern hemisphere owing to the comet's orbital geometry in relation to the Earth and the Sun. Although its initial path carried it across part of the sky well placed for northern observers, this time round the comet, at its brightest, would unfortunately be placed on the far side of the Sun, inconveniently positioned in the daylight sky. Only after perihelion, when the comet's orbital tilt carried it in regions favourable to viewers in southern latitudes (when it would still be totally invisible from the northern temperate latitudes), would it provide something of a showing. Indeed, conditions for viewing from Earth were the worst for two thousand years...

The fact that in tropical and southern latitudes at least the comet *might* provide some kind of "once-in-a-lifetime" event was quickly appreciated by shipping and airline companies alike. This unique situation might be turned to advantage. Most affluent comet spotters were after all resident in the northern temperate zone. If potential comet spotters could be persuaded to travel to the tropics and more distant southern latitudes to view the comet, the adverse viewing conditions might after all prove to be of great commercial benefit.

After Halley's comet was resighted in October 1982, the next prize would fall to the first observer to see it visually through a telescope as opposed to simply recording its tiny image on a photographic plate. In 1909 the first to see it visually had been S W Burnham, using the 101-cm Yerkes refractor, then and still the largest telescope of its type in the world. The fact that Burnham, a double-star observer, had seen it first sorely piqued his observatory colleague E E Barnard, primarily a comet man, and the story goes that Barnard went round with a sour temper for a week after.

In early 1983, Halley's comet was still too faint to be seen except through the very largest telescopes using image intensifiers. On 13 February it was recorded at Kitt Peak as a very faint object shining at about 24.5^m, still almost as faint as when first recovered the previous October. Meantime, while Halley's comet slowly grew brighter and made its run towards the Sun, specialist cometary astronomers had plenty to occupy themselves with. Long expected was Crommelin's periodic comet—so named after the Irish-born computer who had played such a major role in predicting the return of Halley's comet in 1910.

Crommelin, like Halley before him, had not discovered the

comet that now bears his name. In modern times it had first been spotted by Pons in 1818 but then lost after only five days' observation. It was not seen again until independently discovered in 1873 by Coggia at the Marseilles Observatory (Pons's old establishment) and by Winnecke at Strasbourg. Both these men were famous comet hunters of the nineteenth century when rivalry for comet discoveries was at its keenest. The comet was lost again until found in 1928 by an amateur called Forbes in South Africa.

It was at this stage Crommelin enters the story. When he investigated the orbit of the 1928 comet and played comet detective as Halley had done over two centuries previously, he was able to demonstrate that comets 1818 (I), 1873 (VII) and 1928 (III) were the same comet returning at intervals of nearly 28 years. He predicted it would be seen again in 1956.

Like Halley, Crommelin did not live to see the fulfilment of his prediction. By now the comet had become known by the unwieldy name of P/Pons-Coggia-Winnecke-Forbes, and thus it contravened the name rule (*see* Appendix 2). Instead, and in honour of its computer, the International Astronomical Union renamed it P/Crommelin.

On schedule the comet reappeared in 1956, having been found by the famous modern Czech comet hunter Ludmilla Pajdusakova at the Skalnate Pleso Observatory situated high in the Tatra Mountains. However, this time round it was found to be 10° off course. Fortunately after 1956 its orbit was known with much greater precision. In 1983 it was due to return and right on cue it was photographed at Kitt Peak on 9 August 1983 as a dim 20^m object.

Scattered round the heavens at this time were also a score or more of other faint comets which are to be found as visitors to the inner solar system every year. However, like Halley's comet during 1982-4, most of these were exceptionally dim objects and

123 As the 1985–6 apparition of Halley's comet was more favourable to potential viewers in the tropics and the southern hemisphere, travel companies in the northern hemisphere were quick to seize the opportunity for the promotion of luxury comet-spotting cruises. The advertisements were not always accurate in their descriptions of what to expect and these illustrated examples are typical of those offered—including the enticement of drinking champagne while comet spotting!

Cometa

WORLD OF OZ 1986 HALLEY'S COMET EXPEDITIONS

As you read this, Halley's Comet is rushing headlong into our inner Solar System for its rendezvous with the sun in February 1986. Shown below are 5 land programs and 1 sea voyage pr... by our Scientific Expeditions Division. All have been carefully researched and programmed to place tour members in strategic places at the right time for optimum viewing of the comet.

ON LAND

#1 Expedition to South America 18 Days
Departs U.S. Fri. Mar. 28, 1986
Approx. cost **$2995.00** per person
Visits: Rio de Janeiro, Iguassu Falls, Buenos Aires, Santiago, La Serena, Lima, Cuzco and the Inca ruins at Machu Picchu

#2 Expedition to South Africa 16 Days
Departs U.S. Mon. Mar. 31, 1986
Approx. Cost **$2900.00** per person
Visits: Johannesburg, a deluxe private game preserve in Kurger Park, ...etown, an Ostrich Farm at ...n, Durban, a diamond ...ulous multimillion ...al extensions

AT SEA

#6 Special Sun Line OCEANIS Cruise 10 Days
...eparts San Juan
... 4, 1986
...nique
...land

March-April, 1986

See the splendour of Halley's Comet from the roof of Australia.

The crystal clear air at Mt. Kosciusko, the highest point in Australia, will give you an ideal sighting of this exciting event.

Kendell Airlines offers an outstanding travel package to Halley viewers. From Melbourne, you fly to Merimbula on Australia's Pacific Coast, spend 2 nights at the luxurious Black Dolphin Motel, enjoying superb accommodation and sun-drenched beaches. Then on to the majestic Snowy Mountains for 3 days and 2 magic nights at the fabulous Thredbo Alpine Hotel, right at the chairlift to the roof of Australia.

IN PURSUIT OF HALLEY'S COMET.

ONCE IN A LIFETIME EXPEDITIONS. ONCE IN A LIFETIME VIEWS.

Before you know it, the skies of the southern hemisphere will come alive with the most dazzling display of celestial bodies in 76 years. As Halley's Comet rushes toward our Inner Solar System, we at The World of Oz are proud to offer this exciting selection of global expeditions meticulously planned to pl... very best viewin...

experts in astronomy will accom...

CATCH A FALLING STAR.

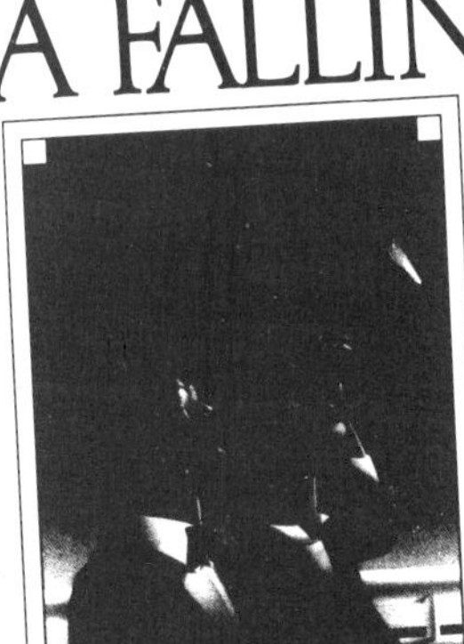

Cheers To Halley's Comet And Champagne Cruising In The Caribbean.

There's no better way to celebrate the romantic Caribbean than Paquet's Champagne Cruising.SM And with the addition of festive theme cruises,* there's no better time than this winter to come aboard. Choose from our dazzling *Halley's Comet* cruises complete with experts from the Southern Cross Astronomical Society. Our *Cousteau Ocean World* cruises with speakers like Jean-Michel Cousteau. Or our *Mystery at Sea, Land of the Maya* and *Wellness* cruises. Even Paquet's world-renowned *Music Festival at Sea*, returning for the 29th spectacular time. Six bubbly themes in all — and every one a champagne occasion indeed. Whether on the French-accented *S/S Rhapsody* or on the *M/S Mermoz* with its classic *atmosphere de France*. Just contact your travel agent for all the heavenly details.

PAQUET FRENCH CRUISES
Ships' Registry: Bahamas.

SAY OUI TO CHAMPAGNE CRUISING.

*Priced the same as our regular Caribbean cruises and include free airfare, free French wine, deckside champagne brunch and award-winning cuisine.
Paquet French Cruises, Dept. ST, 1007 North America Way, Miami, Florida 33132

only recorded on photographs, using the largest telescopes. The re-appearance of Crommelin's comet was keenly anticipated, for in past apparitions it had become a relatively bright telescopic object near perihelion. Therefore, coming as it did just before Halley's comet would predictably grow brighter, it might provide an excellent test-run object for those observers participating in the International Halley Watch, which would be at its peak round 1985-6.

In 1818, Crommelin's comet had been seen as a 7^m object; in 1873 about the same brightness; in 1928 it was 6^m; and in 1956 just over 7^m. This comet, however, had probably been seen in earlier times when it appeared brighter. Apart from linking its former apparitions in 1818, 1873 and 1928, Crommelin had recognized its orbital fingerprints in a comet seen in 1457, the year after the appearance of Halley's comet, and again in 1625. On both these occasions it became a brightish naked-eye object. When Toscanelli recorded it in 1457, it reached 3^m and in 1625 it was even brighter and reached 2^m. During this latter apparition it developed a striking tail over 40° in length.

Although it was obvious that Crommelin's comet had faded from its former splendour, there was always a chance something spectacular might happen. The cometary record contains several instances where faint comets have suddenly undergone a burst of brilliant activity and confounded predictions. However, the record contains *more* instances where bright comets have suddenly faded beyond all recognition. The reason for these unpredictable events provides one of the ongoing mysteries of cometary physics.

Unfortunately, in 1983-4, Crommelin's comet did not quite live up to expectations, reaching only 8^m and just visible in a pair of small binoculars. Nevertheless, in its role as an international test comet in a preamble to check out procedures to be used later for Halley's comet, it proved a great success.

In the early 1984, Encke's comet visited the Sun again, but this time round, like Halley's, it was badly placed for observation. For this reason P/Crommelin was chosen as the test-run comet for the International Halley Watch Programme. Because Halley's comet so dominated the news in January 1986, an anniversary that failed to raise even a single line in the astronomical press was the bicentenary of the first recorded appearance of Encke's comet discovered by the Frenchman Méchain on the evening of 17 January 1786 when on that occasion he reported it plainly visible to the naked eye.

While comet watchers patiently waited for Halley's comet to grow brighter on its long haul towards the Sun, it was subject to a veritable spate of scientific papers—many from astronomers who previously had shown little or no interest in comets. But the difference now was that Halley's was *news* and considered fair game for anyone who wished to carve out an immortal niche for themselves via the journals. If the only way to do this was to cobble up some new theory about the comet or its supposed physical properties, so be it. High on the list of conjectures was the comet's likely rotation period—for it follows that an object with a solid discrete nucleus moving in space would have some kind of spin, or rotation, imparted to it. Because the nucleus of Halley's comet had never been directly observed, its nature and general properties now provided a unique guessing game... Old photographs of the comet taken during its 1909-11 apparition were dug out in their hundreds and "reassessed" or "reprocessed" using computerized techniques. Based on these new interpretations, conflicting rotation periods varying in duration from a few hours to several days were announced by different researchers. Written up in the usual jargon of serious astronomical journals, authors referred to: "gradients and enhancement underlying residual contrast features...along circumferential segments", etc. However, if the authors were honest, their ideas were little more than exercises in inspired guesswork dressed to look otherwise.

In 1983 some of the most exciting news was the discovery of a comet by the recently launched Infra-Red Astronomical Satellite (IRAS). Although this was the first direct automated comet discovery in history, it was also independently discovered visually by two amateurs, the Japanese G Araki and the Englishman veteran comet hunter G E D Alcock. Thus its designation in the comet catalogues became comet IRAS-Araki-Alcock 1983d.

The new comet, apart from the circumstances of its unique discovery, attracted widespread public attention, for on 12 May 1983 it approached the Earth to within 4.6 million km, the closest *known* cometary encounter (disregarding the Tunguska object, *see* note 24) since Lexell's comet in 1770.

For a while the new comet stole the limelight from Halley's. IRAS itself went on to discover a total of *six* new comets in rapid succession (1983d,f,j,k,o,v) before it finally stopped functioning on 21 November that year. The 1980s were to represent a watershed in cometary astronomy with the fly-bys past Halley's comet (*see* below), but the IRAS discoveries also provided a very significant milestone. It heralded a new age of electronic wizardry applied to comets, for in 1982 a team working at the Naval Research Laboratory in Washington DC, using an orbiting satellite, known as Solwind, launched to monitor the Sun, accidentally discovered three new (small) Sungrazing comets which were unobserved from ground level (*see* also note 27). What had once been the subject of a nineteenth-century hoax played on E E Barnard respecting an automated comet spotter (which a newspaper claimed he had invented) was now, at least in principle, 92 years on a reality!*

Yet another event in the cometary world due to IRAS, just six weeks before it stopped operating, was the unique discovery of an unusual object later designated 1983 TB. Because its photograph showed a sharp stellar image, it was assumed to be an asteroid (or minor planet). However, calculation of its orbit showed it moved in a path round the Sun which almost matched that of the Geminid meteors seen round 7-15 December each year and with which previously no known comet had been associated. This raised the question: was 1983 TB simply an extinct core, or nucleus, of a comet?—a comet which, after regularly approaching the Sun to within 0.14 AU, during its short 1.65-year orbit, had now lost all its dust and volatiles? It was an intriguing speculation, for the idea had long been current that some of the Apollo group Earth-crossing asteroids might also be defunct comets rather than mini-planets. On the other hand, could 1983 TB be an

*For an account of this hoax *see* the author's *Comets, Meteorites and Men* (1973/4)

entirely new class of object *halfway* between comets and asteroids? The mystery at present remains unsolved, and 1983 TB is placed high on the priority list of objects to be studied by a space probe at first opportunity.

To observers watching the slowly shifting position of Halley's comet during its three-year run-in from 1982-5, it appeared to trace out some ever-increasing loops in relation to the more distant backcloth of stars. This, however, was only its *apparent* movement brought about by the fact the Earth is continually moving in its annual orbit round the Sun. Because of its great distance, Halley's comet remained in the same general region of sky in Canis Major and Monoceros, between Gemini and Orion, from October 1982 until the end of 1984. By 1985, with the comet now much closer and accelerating, its apparent path carried it successively over Taurus, Aries and Pisces—the region of sky from where it would be widely observed during the latter part of 1985.

Information on the brightness of the comet during 1983 was scant and some of it contradictory due to occasional variations in brightness reminiscent of those noted by Barnard and others back in 1909-11. By November 1984 it was still moving against the backcloth of stars between Gemini and Orion and it would remain in this region for some time to come. It now shone at 18^m and lay at a distance of 5.7 AU. At this time it was about 72 million km farther from the Sun than Jupiter (Fig 91).

The comet's regular fluctuations in brightness, sometimes by as much as $1\frac{3}{4}$ magnitudes, now led to some diverse estimates for the rotation period of the comet's nucleus, ranging from 8 to 48 hours.

In September 1984 the comet had been photographed by the first amateur astronomer to record it. This was the Japanese observer Tsutomu Seki, joint discoverer of the brilliant Sun-grazing comet of 1965. Using his 60-cm reflector, Seki estimated its brightness to be 20^m, and this figure generally agreed with the estimates made by professional astronomers at the Kitt Peak Observatory using image intensifiers.

By January 1985 the comet had moved in to a distance of 5 AU, or just slightly closer to the Sun than Jupiter. It was now that an astronomer named S T O'Meara claimed to have sighted it *visually* on 22 January using a 60-cm reflecting telescope located on the high summit of Mauna Kea in the Hawaiian Islands. At this time he estimated its brightness to be 19.6^m. Never before had any astronomer claimed to have identified a comet when it was so dim with such a relatively small aperture telescope. Subsequently this observation has not met with universal acceptance. Some astronomers believed O'Meara could not possibly have identified the comet among the myriad faint field-stars and that in his enthusiasm to be the first to recognize it visually he had overreached himself. If he did see it, he was the first astronomer to do so directly since E E Barnard had last glimpsed the retreating comet with the 101-cm Yerkes refractor on 23 May 1911 when he had described it "as exceedingly faint and diffuse". The question whether O'Meara actually saw the comet still remains an open one.

During this period the comet was also observed by astronomers utilizing alternative modern techniques. In December 1984, British astronomers using radar (also operating from the summit of Mauna Kea) announced their measures indicated that the nucleus was at least 15 km along its largest dimension. This indeed proved to be a very close estimate and compared favourably with measures later determined by the Soviet Vega and European Giotto probes when they encountered the comet in early March 1986 (Chapter 20).

During April 1985 the comet crossed the 4 AU distance mark, but observers noted the coma region was still very small. By this time the comet had brightened to 16^m.

By June the comet was lost in the daylight sky; at the end of the month it had closed to a distance of $3\frac{1}{2}$ AU and was moving between the orbits of Jupiter and Mars.

By August it was a morning object in the eastern sky, rising just before twilight. The comet had remained hidden in the Sun's glare for several weeks and during this time it had slowly brightened and expanded, transforming its appearance into a fuzzy patch of light. It was still 480 million km from the Sun, yet even at this great distance the action of the Sun's radiation was already beginning to sublimate the hidden nucleus.

In July, while the comet was still invisible, there was strong competition between amateurs and professionals to be the first to announce its reappearance. By 28 July several amateurs with large instruments were reporting to have seen it; however, the earliest known recovery was by Richard West (of comet West fame) who photographed it on 19 July from the European Southern Observatory in Chile when it revealed itself as a 16^m object.

Somewhat disappointingly the comet was now about one magnitude fainter than forecasts had indicated it might be.

By early October the comet had crept to within 2.01 AU from the Earth and 2.33 AU from the Sun and was very well placed for general observation. It was round this period that world-wide some tens of thousands of professional and amateur astronomers, armed with more modest equipments, made their first direct acquaintance with the venerated object that had for so long occupied their thoughts.

By 26 October it had brightened to 8.5^m—a little brighter than predictions for it—and suddenly there was a new note of optimism that the comet after all might ultimately put on something of a public display against all earlier and recent expectations. It now appeared as a small cloud 8′ in diameter (a little less than a quarter of the Moon's apparent size), representing a true diameter of some 480,000 km. However, this did not include the vast tenuous but totally invisible hydrogen cloud surrounding the comet and extending to a million km—similar to the clouds first detected round bright comets seen in the late 60s and 70s.

On 29 October the comet passed within less than $\frac{1}{2}°$ south of the famous Crab Nebula pulsar (M 1) and, in spite of a bright Moon, it was easily detected, appearing brighter than the nebula itself rated at 8.4^m.

By early November the optimism engendered in October had so far come to nought. Unfortunately these new expectations had been passed on by the media to the public without qualification.

Although it was all too plain that the comet was far from being any kind of dramatic public spectacle and many would-be comet spotters claimed great difficulty in finding it, comet fever in

Europe and North America now reached a peak, and in some quarters Halleymania prevailed...

Halleymania was certainly no new phenomenon. Back in 1910 the comet was used to promote products ranging from custard powder and fountain-pens to corsets and cosmetics, and the inevitable cranks predicted the end of the world. However, the outbreak earlier in the century was a comparatively mild form compared with the new strain of the mid-1980s.

Immediately after October 1982, Halleymania had been slow to show itself. This was not surprising; after all the comet was still three-and-a-half years away from its perihelion, and not even the most optimistic were expecting much of a display before autumn 1985.

Behind the scenes in the early 1980s the plots were already thickening, but it was not until late 84/early 85 that a real quickening of interest was readily detected. By the time the *Wall Street Journal*, a publication previously unknown for its comet reportage, carried a prominent leading article on the rapidly burgeoning commercialization of the predicted great event, it was abundantly clear that the hucksters were suddenly in full cry and advertising hype the order of the day.

During 1984-5 there spawned a diversity of comet souvenirs ranging from expensive warm-up jackets to throwaway breakfast cereal packaging. On sale in New York stores in January 1985 was a choice of at least sixteen brands of Halley T-shirts. But this was only the start of things, and by 9 February 1986 the number of comet products on offer world-wide numbered many thousands.

One conservative estimate rated the Halley market to be worth at least \$500 million. In attempts to hog the market, several entrepreneurs claimed the word "official" in their advertising handouts. Even the authors of down-market comet books were bent on similar mischief. One British self-publicist organized himself as head panjandrum of a quasi-masonic "Halley Society" which had all the fin de siècle overtones of a Gilbert and Sullivan

124 Halleymania 1985–6. Among the myriad of promotions were comet-motif T-shirts and, world-wide, several hundred variations-on-a-theme were on sale. Some examples are illustrated here, along with a special comet hat for the occasion.

comic opera. It adjured its membership to promise faithfully to enunciate Halley's name in accordance with the archaic Hawley form. Fortunately for good sense, this archaic form (*see* note 1) was to be totally ignored by the BBC's Pronunciation Unit and everyone else.

When Seneca recognized human fallibility in respect to mankind and comets nearly two thousand years ago, he perhaps hoped that in the distant future we might be cured of it. Yet the 1980s showed little change in human nature. When one get-rich-quick entrepreneur hatched a scheme by offering shares in Halley's comet, incorporated "under the laws of the Milky way" and not redeemable till 29 April 2061, there were credulous subscribers eager to part with their money.

Another instituted a "Love the Comet" fund which also found a ready subscription list. One brazen huckster leased out "under licence" the Halley's comet logo copied by him straight from the thousand-year-old Bayeux Tapestry! When several companies willingly paid him handsome retainers for the privilege of using it to label their Halley products, this again underlined mankind's universal gullibility.

Inevitably comet pills made their appearance. One of these was marketed openly as a joke product to celebrate "75 years of human gullibility" and took the form of a yogurt-coated sunflower seed. Bottles of these pills retailed in US at $2.95, and there were many takers.

Outbreaks of comet fever in its Halleymania form were not confined to Anglo-Saxon races or temperate-clime countries as this pandemic respected no frontiers. A *New Scientist* correspondent visiting Brazil reported its effects there were as virulent as any witnessed in more northern zones.

In spite of the fact that the name Halley proved as unpronounceable in Portuguese as in Japanese, Halley hype had no difficulty in surmounting this potential anti-commercial hazard. One astute Brazilian huckster, taking an early-bird cue from his more northerly brothers, succeeded in registering "Halley" as his personal trade name. The result was that manufacturers pushing diverse products—ranging from dolls brand-named Halleyuiah and Halleyfante to tennis rackets bearing the Halley logo—were

Daily Mail, Saturday, January 18, 1986

Britain crows over its star turn . . .

Strange tail of the Comet Egg that beats the world

By JOHN PASSMORE

LAST TIME it happened in Reno, Nevada. Before that it was Marburg in Germany. This time the Halley's Comet Egg has turned up in England.

Regular as clockwork, every 76 years when the wandering comet visits the earth, it is said that a hen somewhere lays a 'Comet Egg'. With a £5,000 prize at stake, British hens alone produced no fewer than 350 this time around. There were eggs with star shapes, eggs with speckled tails, eggs with lumps and ridges.

But the one Linda Franklin nearly threw away at Becketts Poultry Farm in Studley, Warwickshire, was the oddest of the lot.

Regional Egg Inspector Kim Miller from the Ministry of Agriculture reverently picked up Mrs Franklin's winning egg from its tray yesterday, frowned, shook his head in disbelief and said: 'I've seen eggs by the millions with every sort of imperfection—but I've never seen anything like this.'

Mrs Franklin, 38, took the egg home during a spell working for the poultry farm, to show her 12-year-old son Tony. 'I didn't know anything about comet eggs' she said. 'I never even saw the comet. It was my mother who told me about the competition.'

But then legends have a habit of working themselves out against all the odds. A 300-year-old woodcut shows that the Marburg egg was marked with a five-pointed star.

And in 1910, County Clerk Fogg of Reno went out into his backyard to see the comet and found his pet hen had laid 'an egg with a long tail on it'.

Frank Powell, marketing director of Thames Valley Eggs, who organised the egg hunt, announced: 'We'll be offering it to a museum.'

The Egg and I . . .

STAR-STRUCK HEN COULD LAY £10,000

Thames Valley Eggs is offering a £10,000 prize for the first egg to be laid in Britain during the next two months with a comet-shaped marking on its shell.

The Berkshire company says there are records of such eggs being laid during previous visits by Halley's Comet.

Mr David Watts, managing director, said the prize egg would have to show a star with a tail etched in the natural pigmentation of the shell. The owner of the hen producing it would get only half the prize, the other half going to the Duke of Edinburgh's Award Scheme.

125 Halleymania 1985-6. As anticipated, there were echoes of the original comet-egg episode circa 1680-1 (*see* Fig 16).

required to cough up royalties varying from 7 to 30 per cent.

At least 400 Halley products flooded the local market, and popular magazines and newspapers were filled with comet advertising.

A local feature, which H G Wells might have inspired, was Brazil's official "Halley Day of Harmony" declared on 11 April 1986.* Alas, unlike Wells's famous science-fiction comet, circa 1906, its influence on mankind was only transitory, and any noticeable "gas effect" was confined to the utterances of human banalities via the domestic radio coverage.

The comet also perhaps exampled one of the differences between Latin and more Germanic temperaments. Whereas in Britain and elsewhere there was a vogue in the public display of self-parodying lapel-badges with the message "I've seen Halley's Comet", in Brazil it seemed nothing short of a self-assuring

*The day the comet was closest to the Earth (0.42 AU) on its outward leg.

signed document for domestic framing and pride-of-place display, containing much the same message, was considered worthy enough for this great event in history.

Newspapers and the media generally, initially keen to exploit the Halley phenomenon for all it was worth—recognizing it as a source to fill their columns, air and viewing time over several months—showed a remarkable cooling off as the apparition progressed and the promises by some not-so-truthful self-publicists failed to materialize. "It's comet Kohoutek all over again!" claimed one disgruntled columnist. Another, in Chicago, disparaged it as "this gaseous fraud".

On 31 December 1985 one British newspaper devoted the whole of its editorial leader to discussing the public's general disillusionment, remarking: "...The name Edmond Halley invites interest and admiration for scientific achievement, but these are not the emotions that *a comet* is supposed to excite." And because the comet was still quite beyond range of the naked eye to the average citizen viewer, the writer added wryly: "... there are not usually many people around at night carrying binoculars and there is perhaps a feeling that a real comet is one that can be seen clearly enough with the naked eye for instructions about discovering it to be unnecessary."

Indeed most astronomers as well as the general public shared the same opinion.

But it was not only in Britain and the rest of Europe where the comet had been disappointing. Across the Atlantic, the Mayor of New York promised the many frustrated Manhattan citizens that as an aid to seeing it he would switch off all the city lights on the first suitable occasion...

Media coverage was generally ill informed. Among common misconceptions was Halley's title. Frequently on TV and in newspapers he was elevated to "Sir Edmond Halley".

Even those programmes purporting to tell the *scientific* story of Halley and his comet were often riddled with errors of fact. One, very badly researched, hour-long British programme claimed that in 1910 Crommelin and Cowell were in error by 12 days. It was implied they were really duffers at the job compared

Daily Mail, Tuesday, November 12, 1985

Exposed at last! Pathetic tail of Halley's comet

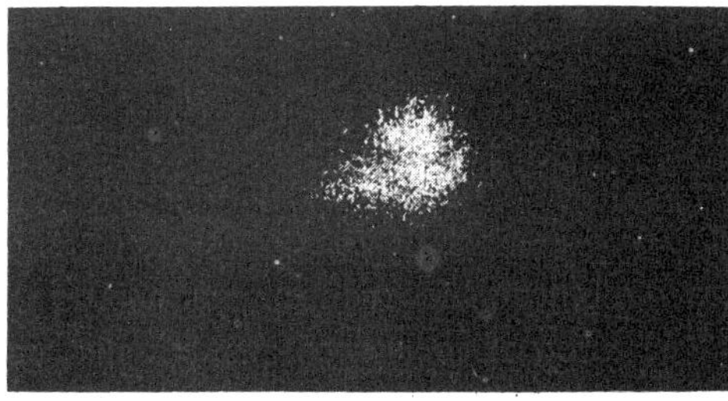

THIS is the photograph of a lifetime — the first clear picture of Halley's comet trailing the most famous tail in astronomy.

It looked like a glowing tadpole as it hurtled through space 80 million miles from earth.

But the tail that has inspired poetsand prophets was little more than a blob.

The picture was taken by Mailman Peter Lomas at the Lancashire Polytechnic and he used a Nikon 35mm attached to a 15-inch telescope to get the snap.

Well, not quite a snap. The comet was tracked through the night-sky for one hour and twenty minutes exposure time to ge a clean image.

The photograph was greeted with mixed feelings yesterday by scientists at the Polytechnic who have spent months preparing equipment.

Astronomer Dr Ian Butchart said: 'Ths is the first decent picture we have had of Halley's comet and obviously we are delighted.'

And the tail that has been blamed through history for death and disaster, and traditionally has been portrayed as a spectacular brilliant light show ?

'It's a bit disappointing,' said Dr Butchart. 'The tail is really quite pathetic.'

As part of National Astronomy week the Polytechnic's observatories will be open to the public each evening so they can view the comet

126 When the British mounted a special astronomy week rather prematurely in early November 1985 so as to awaken and encourage public interest in Halley's comet, then still below naked-eye visibility, the event received a mixed reception. Many of the general public invited to local observatories to view the comet came away disappointed at its miserable showing, and this was reflected in newspaper comment. It was shades of 1910 all over again...

... Nevertheless, there were some enthusiasts determined to make it the event of a lifetime.

Comet buff hires jet airliner

BUILDER David Jupp explained why he is chartering a jet airliner to make sure he gets a glimpse of Halley's Comet. He's bored.

"Every morning I get up. I go to work. I do my job," he said.

"Every evening I come home again. I've been doing it for 26 years. "Christmas, Easter, summer holidays — they're all the same to me. This is something you only do once, and that's why I'm doing it. For no other reason."

Mr. Jupp, 41, of Edgware, North West London, is paying £3,800 to charter a Britannia Airways Boeing 737, and selling tickets at £38 a time for those who want to share the experience.

He tried through his travel agent friend Gordon Buist to book seats on an official "Comet flight" but discovered they were all booked solid with would-be stargazers.

127 As the comet drew closer to the Sun during the latter part of 1985 and early 1986, the media appropriated the comet from the astronomers and added to the extravagant hype already engendered in some quarters... Even Schultz couldn't resist a mention of Halley's comet in his lovable "Peanuts" cartoon. Fortunately the information imparted to his world-wide audience was a good deal more accurate than most of that emerging from self-acclaimed TV pundits.

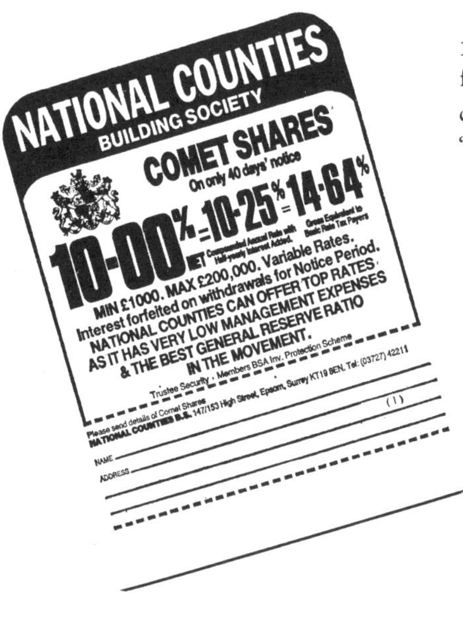

128 In 1985–6 even financial houses felt obliged to cash in on some of the comet's publicity factor and advertise "Comet Shares".

with latter-day American computers. Hind's efforts, as well as the toils of Clairaut, Lalande and Lepaute or any other computer before 1910, did not rate a single mention. The idea was implanted it was all due to present-day (US-based) computers that the comet had been traced back in history, first to 1066 and then through Chinese records to 240 BC. This programme was blithely put out again a few nights before the Giotto probe made its historic encounter without the least attempt to correct the inaccuracies.

A thorny subject aired in astronomical columns was *when* the comet actually became known as Halley's comet. Several historians of science claimed it was the French astronomer N L De Lacaille (1713-62) who first coined the name in May 1759. However, this is in error by more than half a century, for both David Gregory and William Whiston made reference to "Mr Halley's Comet" or "Captain Halley's Comet" on several occasions immediately post 1705 after publication of the *Synopsis*.

There were also some wild misconceptions aired by journalists about Halley's precise role in linking the comets of 1682, 1607 and 1531. One naive TV journalist suggested that Halley was really an opportunist (which perhaps he was in other contexts), for had not the Chinese, he insisted, been observing this same comet long before Halley was born, and, therefore, how could he appropriate it for himself without Newton and others raising objections to this! Indeed, he believed he had hit upon a great truth not till then obvious to others. I leave the reader to ponder the flaw in his argument.

Halley's comet in 1986 also indirectly provided a telling document for future social historians. When in 2061 the comet comes round again, text-books will no doubt imply how by 1986 the bad old days of poverty and under-privilege had disappeared, and that when the comet last visited, Britain was an affluent democratic society. A glance at the archives, however, might cause some to reflect on the verity of this. If they consult the *Daily Mail* for 3 March 1986, they will read a report of how the hard winter of 1986 was killing off hundreds of pensioners every week simply because they could not afford to buy fuel for domestic heating. For contrast, they should then turn a few pages on.

Here they will read how a privileged few could afford to indulge themselves to the tune of £7,795 each on a capricious jaunt to the southern hemisphere aboard Concord for a quick look at the retreating comet.

During November 1985, as the comet's distance from the Sun decreased from 1.9 to 1.5 AU, it moved across the constellations of Taurus, Aries and Pisces. It was now well within range of small binoculars and could be followed all night long.

On 16-17 November the comet was uniquely placed when it passed 2° south of the famous Pleiades star cluster. In binoculars both objects were in the same field of view, and this allowed the comet to be spotted easily by the uninitiated.

The night of 27 November marked the closest approach of the comet to the Earth on its inward journey to the Sun. At this time it lay at 0.62 AU, or roughly a distance of 93 million km. Compared with the comet's nearest approach to the Earth in 1910, this was still a relatively large distance. Any good view of the comet, however, was spoilt by the presence of the overpowering light of the full Moon, making it impractical or difficult for the general public to find it. Many disappointed viewers subsequently echoed Sam Pepys's difficulties when the comet of 1664 first eluded him in bright moonlight.

During the latter part of 1985 and early 1986 either the Moon or street lights, or both, were a general discouragement to widespread public viewing. But, in locales where the night sky was dark and the air transparent, naked-eye sightings were reported by the second week in November. Yet, even under the most ideal conditions the comet, as a naked-eye object, was only to be seen as a faint fuzzy star at the limit of visibility.

In early December, now nearer the Sun, it developed a 2° gas tail; then in the following weeks a distinctive, broader dust tail also gradually developed. Although still a run-of-the-mill object, in comparatively modest telescopes the coma region was very active, and there were fine displays of jets, fountains and rays

129 By early January the Halley 'tadpole' had developed a longer tail and *through larger telescopes* began to provide a more impressive spectacle as depicted here in an enlarged photograph taken by the Soviet 1-m telescope at Mt Sanglok.

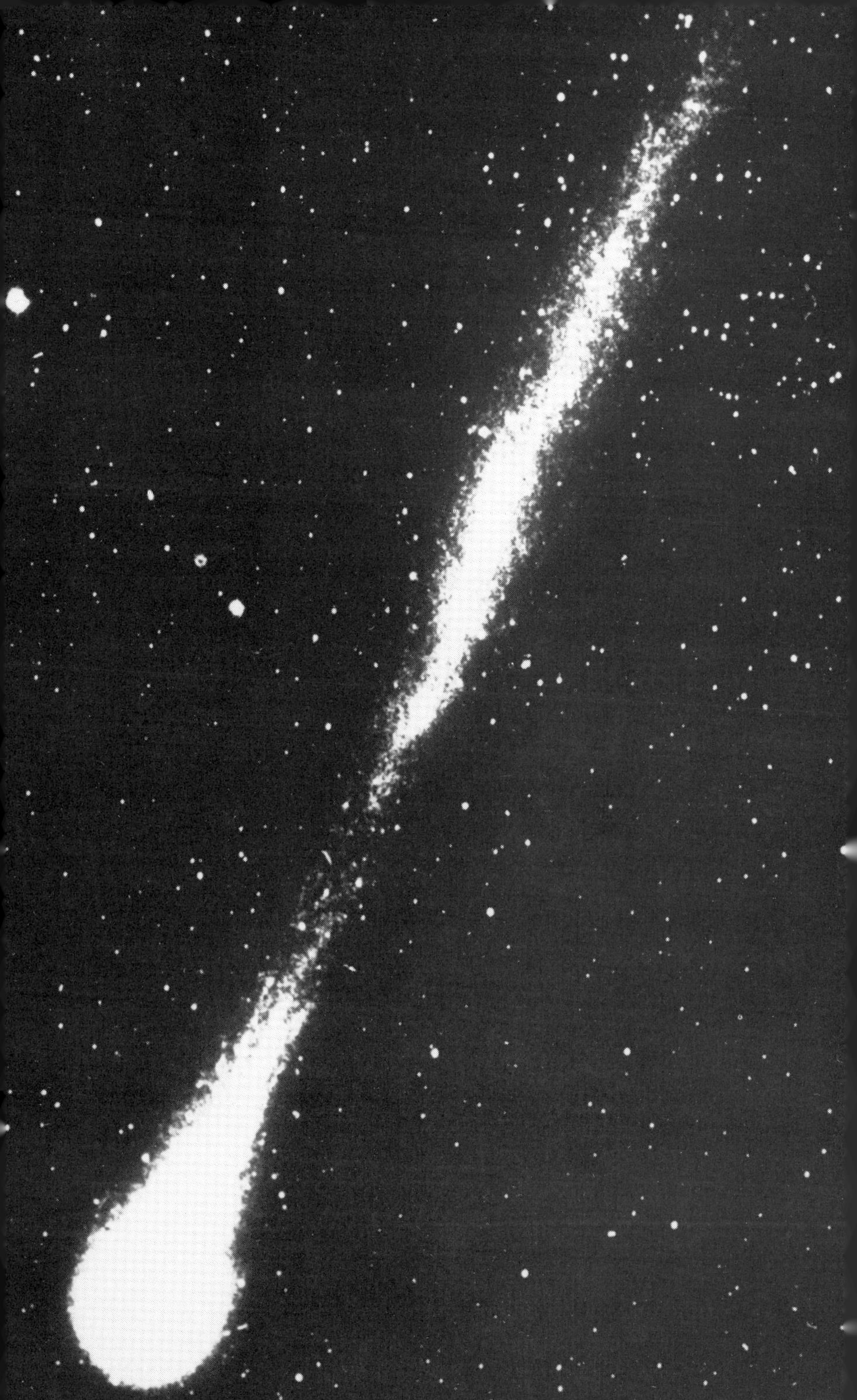

emerging from the head like those noted and drawn by observers in the past.

Any hopes for a bright Christmas comet were dashed when the comet failed to rise in brightness much above 5^m. By now the optimistic note engendered in late October withered to stoical pessimism. During January 1986, when the comet was expected to put on a brighter display as it moved sunwards towards its 9 February perihelion and failed to do so, newspaper editorials were again critical.

By late January the comet, having shown reluctance to brighten further, was already lost in the twilight sky. During early February, at the time of its perihelion, it remained totally invisible from Earth. While newspaper editors and the general public had been totally unimpressed by the comet's performance so far during its apparition, astronomers stayed optimistic. The comet's reappearance, round mid-February, would herald the final count-down to the most exciting phase of the international observing programme.

CHAPTER 20

Close Encounters: A Glimpse of Genesis?

"But who shall dwell in these worlds, if they are inhabited?"

—KEPLER

"The further we go and the more closely we study, the more plainly it is brought home to us that we are waifs shipwrecked on the ocean of nature; and ever and anon, from a sudden wave that shall be more transparent than others, there leaps forth a fact that in an instant confounds all we imagined we knew."

—MAETERLINCK

In the mid-1960s, after the new space technology became available and probes to comets were first seriously mooted, no one could then foresee that two decades would pass before the cherished dream was realized. Halley's comet *was* the ultimate target, but in the interim several dress rehearsals were deemed necessary with probes to other comets to learn the lessons for the mid-1980s. Among those in the offing was Encke's comet. With its short $3\frac{1}{3}$-year period and direct orbit, this was a relatively easy target compared with the complicated retrograde-moving comets like Halley's. However, no one foresaw the big cut-backs in international space programmes and the reallocation of priorities that were to follow in the 1970s.

Then in the early 1980s, after nearly two decades of neglect, it came as a mild shock to discover suddenly that the long-delayed new age for cometary science was dawning and there were at least five probes planned for Halley's comet with no chance for dress rehearsals (Table 3). If all these came to fruition, then Halley's comet was going to be the focus of world-wide attention when a small armada of spacecraft made their rendezvous with it round March 1986.

What helped break the stalemate and change priorities was

Halley's comet itself. Halley's was the recognized big name in comets and part of world folklore. Even the man in the street had heard of it, and it was appreciated that those agencies with a probe going for the comet would attract an unusual proportion of international publicity and commercial kudos. Indeed, British Aerospace, who built the Giotto craft for the European Space Agency, used this to great advantage when they went public and launched their shares on the stock market in 1985.

Nevertheless, all these probes to the comet were much more than publicity exercises, for in spite of all the high-tech electronic equipment available, it was universally appreciated that any radical break-through in knowledge about comets via ground-based observatories was unlikely ever to be realized. While the orbiting laboratory-observatories of the late 1960s and early 70s (p 219) had pioneered new methods and opened up new vistas for observing comets immediately outside the constraints of the Earth's atmosphere, even these were limited in scope. If a probe could actually fly close by a comet and pass through its outer coma, Seneca's still unrealized prediction that man would one day solve the riddle of the comets might at last be fulfilled.

First of the major launches were those by the USSR. On 15 December 1984, Vega-1 was successfully placed in orbit, then six days later Vega-2 followed in its wake. Originally both these probes were planned only for Venus as part of the very successful programme for the exploration of this planet conducted by the Soviet Union over many years. Fortuitously, however, Soviet planners suddenly realized that each of these two Venus probes could be rescheduled through a launch window to provide for a very unique double event—first Venus in 1985 and then Halley's comet in 1986...

The name Vega stems from the Russian contraction of Venus-Halley (Venera-Gallei, the letter H becoming an aspirated G). Like the European Space Agency's Giotto mission, the Vega project drew on technology from several countries. Those involved

130 Comet fever of another kind: A stock market promotional advertisement, to woo investors, placed in a newspaper in 1985 by British Aerospace illustrating the European Space Agency's Giotto probe to Halley's comet for which they were one of the main contractors. (above) City editor's comment.

Daily Mail, Saturday, April 13, 1985

City Investment Extra

All aboard Giotto to catch a comet

By Michael Walters

AS HALLEY'S Comet flashes past Planet Earth at 35,000 miles an hour next March, John Hartley will be holding his breath.

And he will be hanging on every signal from a tiny spacecraft zooming towards the cosmic dust cloud behind the Comet at a closing speed of 150,000 miles an hour.

Halley's Comet comes close to earth but once every 76 years. This time round space teams from Europe, Russia and Japan will be shooting at its tail hoping the dust and data they can gather from it might help unlock little secrets like the origins of our planet.

The European satellite will be called *Giotto*, after the Italian painter who pictured the Comet in the background of his painting *The Adoration of the Magi*. The ship will be built by British Aerospace, and Mr. Hartley's company, International Signal and Control, makes the computer which controls the on-board computer systems.

Mr Har ... 'It's eerie alr ... The con ... runs the ... but che ... OK its ... left ... say, ... can ... with ...

R ... IS ... Sp ... a ... r ...

since 1966, space is only about one eighth of ISC's business. Most of the turnover is in highly sophisticated electronics and defence systems. They do not talk about it too much, but things like the Rockeye cluster bomb are ... big business for ISC. ... American company ... raised capi- ... nd have ... gh

East blow-up could threaten business, and there are dangers that big customers could default on their payments. But in 1981, ISC made pre-tax profits of $5m. They probably made $39m. for the year which ended on March 31. This year, they could make $50m.

High risks bring high rewards. Since I first high-lighted ISC's attractions four years ago (they were then called ESIL), the shares have multiplied 17 fold. Since I returned to them three years ago, they have multiplied more than four-fold.

Now at 275p, the shares sell on about 13 times ... spective 1985-86 earn- ... will not go up ... mes in the ... but

PAGE 29

Daily Mail, Wednesday, April 3, 1985

This time round, Mr. Halley, we're going to catch your comet by the tail.

Last time Halley's comet put in an appearance, space flight was no more than a gleam in H. G. Wells' eye.

This time round it's a reality, and at British Aerospace we're leading the design and construction of Giotto, a scientific research satellite that will intercept Halley's comet when it approaches us next year.

British Aerospace, under contract to the European Space Agency, is heading a multi-national consortium in the design, development and manufacture of the Giotto spacecraft and its associated support systems.

Giotto, scheduled for launch in July, is just one example of the programmes that make British Aerospace a world leader.

British Aerospace is one of Britain's largest exporters of manufactured goods. Last year over 60% of the company's total sales were to overseas customers.

We are involved in designing and building civil and military aircraft with sales to over 40 nations. We design, produce and sell a wide range of guided weapons systems, communications and scientific spacecraft.

Our commitment to the future shows itself constantly in innovative design, backed by advanced production techniques and research facilities. All to ensure that we remain in the forefront of aviation, electronics, space and weapons systems technology.

In January it was announced that an offer of shares in the Company was proposed. If you would like to find out more about us, send for a copy of our illustrated Fact Sheet.

To: British Aerospace, P.O. Box 500, Bristol BS99 1BA.
Please send me a copy of your latest illustrated Fact Sheet.
Name (Mr/Mrs/Miss) ___
Address ___
Postcode ___

The world looks up to British Aerospace.

Issued by Kleinwort, Benson Limited on behalf of British Aerospace Public Limited Company and by Lazard Brothers & Co. Limited on behalf of H M Government.

included Austria, France, West Germany, Hungary, Poland, Czechoslovakia, Bulgaria and East Germany.

Each of the Vega probes was programmed to approach Venus and then, via a launcher-craft attached to parachutes, release a 4-metre diameter balloon that would drift in the sulphuric-acid-rich Venusian atmosphere and transmit meteorological data back to Earth.

Without hitch both probes reached Venus in June 1985. It was now as they successively passed at distances of 30,000 km the mission planners took advantage of the planet's gravity field and retargeted the main probes on new courses for their rendezvous with Halley's comet some nine months later.

Both Vegas were armed with a battery of equipment to make the best of their fleeting encounters when, owing to the comet's retrograde orbit, the probes and the comet would be travelling in opposite directions. Among the 15 experiments were those for decoding the comet's gas and dust content, its spectra polarimetry, its magnetic field and various studies at infra-red wave-

131 Trajectories of the Soviet space probes Vega 1 and 2 to Halley's comet. These probes were also coupled with probes to Venus. After a successful rendezvous with Venus in mid-June 1985, both vehicles made use of the planet's gravity-field to redirect them and set them on intercept courses with the comet. (1) Launched from Earth mid-December 1984. (2) Venus encounters mid-June 1985. (3) Encounters with Halley's comet early March 1986.

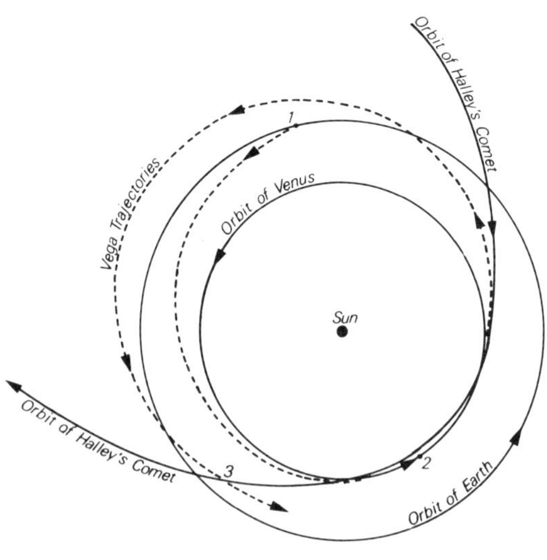

lengths. Onboard dust counters would also transmit directly the impact of any cometary particles as each probe homed in towards the coma. However, chief interest lay in the two platform cameras, mounted in each probe, which could be swivelled by the mission controllers to follow the comet at time of rendezvous. If Vega-1 achieved its target at an approximate range of 10,000 km, as planned, it was expected to resolve physical detail on the comet's nucleus as fine as 200 metres across; if Vega-2 could be manoeuvred even closer, to 3,000 km, the resolution could be greatly increased.

On 8 January 1985, Japan launched the first of its small drum-shaped probes towards the comet. This was named Sakigake ("Pioneer"—formerly designated MS-T5) and was principally a test launch to act as full dress rehearsal for Japan's main probe Suisei ("Comet"—formerly designated Planet A), which was later successfully launched on 18 August 1985. Sakigake would never approach Halley's comet closer than several million km, and its main task would be to study the effects of the solar wind near the comet.

The second probe, Suisei, was targeted to fly by Halley's comet at a distance of 145,000 km. Its aims were to study and map out the vast tenuous hydrogen gas cloud surrounding the comet and also to provide data on the rate of erosion in the comet in order to calculate how much of its mass might be lost during each revolution round the Sun.

By far the most ambitious of all the international probes was the one-shot Giotto mission mounted by the European Space Agency (ESA) and successfully launched from Kourou in French Guiana via an Ariane rocket on 2 July 1985.

The Giotto mission was so named following the supposed identification of Halley's comet on Giotto's fresco the "Adoration of the Magi", contained in the Scrovegni Chapel in Padua, known to have been executed soon after 1301. However, as already pointed out (p 184), this identification with Halley's comet seems an unlikely one. In my opinion the second bright comet of 1301, seen just before Christmas (Fig 87 b), offers itself as a more likely candidate for Giotto's model.

Giotto was scheduled for a seven-month cruise before it

reached Halley's comet on 13 March 1986 (shortly after the Soviet probes), a little over one month after perihelion. At this time the comet would be 0.98 AU distant from the Sun and 1 AU distant from the Earth. If all went well, it would pass within 500 km of the comet during which time it would be moving at 68 km per second. Because of the planned very close approach, there was a real possibility of the probe being destroyed or disabled when it encountered the heavier dust particles surrounding the comet, and for this reason its systems, including its cameras, were designed to relay all their data back to Earth immediately rather than store it for later transmission.

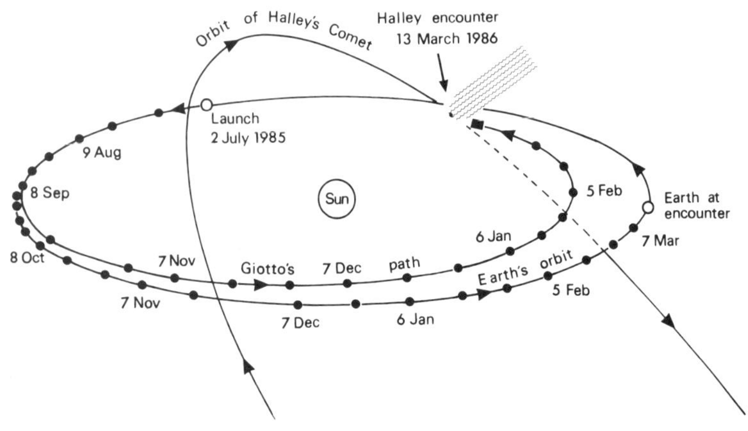

132 The trajectory of the Giotto space probe to Halley's comet in 1985–6.

It was clear that the Giotto probe, because of its ambitious close approach and versatile equipment package, consisting of 11 experiments, might be one which would reveal most; this was assuming that the dust surrounding the comet did not destroy or disable it before it had chance to transmit to Earth any close-up pictures. If the probe succeeded in reaching its target distance, it was expected to reveal detail on the comet's nucleus down to a resolution of a few tens of metres.

On 11 September 1985, while the Japanese, Soviet and European probes were still on course to Halley's comet, a new page in space history was written when NASA, using its International

133 The relative size of the Giotto spacecraft seen during factory assembly.

Comet Explorer (ICE), made a successful fly-by of comet Giacobini-Zinner. This periodic comet, first discovered in 1900 by Giacobini at the Nice Observatory and then independently by Zinner at Bamberg on its 1913 return, was a familiar returnee (with a period of 6.4 years) to the inner solar system.

The ICE probe had in fact never left the Earth under this name. It owed its existence to a former probe named Sun-Earth Explorer (ISEE-3), first launched in August 1978 to study the solar wind, and its historical encounter in 1985 with P/ Giacobini-Zinner was the fortuitous result of some mission replanning on the part of NASA controllers.

When NASA mission planners discovered to their surprise that their old satellite had sufficient power left, they rescheduled it to encounter the comet and manoeuvred it through its tail some 10,000 km from the head. Even after its successful Giacobini-Zinner mission, life in the old probe was far from finished, and it was retargeted again, this time towards Halley's comet so as to pass within 30 million km of it in early March 1986...

Thus in spite of past disappointments and setbacks for NASA, due largely to budget constraints, they were now able to claim the first artifactual encounter with a comet. Although little was revealed about the nature of Giacobini-Zinner's nucleus, the probe collected a rich store of data. In particular it shed light on the action of the solar wind and the bow wave imparted when the comet ploughed through it. It was all very useful scientific grist for the mill for those with similar experiments aboard the other probes heading for Halley.

Another old NASA satellite brought to bear on Halley's comet was Pioneer 12 (PVO) launched on 20 May 1978, which since December 1978 had been orbiting Venus. This was now included as part of NASA's programme to observe Halley's comet near the time of its perihelion in early February 1986 in conjunction with manned observations made from the NASA operated Earth-orbiting shuttles.

While it was appreciated that observations of the comet from orbiting Earth/Venus satellites were limited in scope, as back-up observatories to the main probes they might fulfil an important role. The Skylab crew had made some very successful observations of the ill-fated Kohoutek comet in 1974, and in 1985-6 it was anticipated that the manned US shuttle flights would contribute some important functions in observing the comet during the weeks either side of its perihelion when it would be difficult to observe the comet from ground stations.

The US shuttle flights were practically routine operations by now. Although there had been some disappointments and setbacks (owing to subsequent malfunctioning satellites launched by the various shuttles), these shuttles were considered prime tools for extra-terrestrial research.

On its January 1986 flight the shuttle Challenger, scheduled as the 25th flight in the programme, was due to release a free-floating instrument package that would drift in space for up to 45 hours during which time it would aim two ultra-violet spectrometers at Halley's comet from a distance of 160 million km. Six weeks later the Columbia shuttle would carry aloft a pay-load of equipment (valued at $55 million) called Astro-1 consisting of three ultra-violet telescopes and two wide-angle cameras. All these instruments would then be targeted at Halley's comet to coincide with those days when the five international probes would be making their respective rendezvous with the comet.

The late-January 1986 Challenger launch was initially dogged and delayed by several equipment faults and inclement weather conditions. Then, on 28 January, disaster overtook it when shortly after launch it exploded, killing its seven astronaut crew. From that moment on the shuttles were out of the Halley's comet programme as all further launches were delayed for at least a year.

On 9 February 1986, Halley's comet, now at its perihelion and lost from view from Earth, passed round the Sun 88 million km distant from it. During this time the shocked world reflected on the Challenger tragedy while its crew and equipment carried aboard to observe Halley's comet lay unrecovered at the bottom of the North Atlantic off the Kennedy Space Centre.

After perihelion the comet moved into the morning sky, and now there was keen competition to be first to spot it just before Sun up.

Among the earliest was Richard West, who on 20 February, situated in the very favourable geographical location of the

Southern European Observatory in Chile, reported the re-emerging comet provided a dramatic sight now sporting at least seven tails. For several days following this report there was a new optimism that the comet would become brighter than the 3.6^m forecast for round this time. Once again, however, the comet failed to live up to these expectations, and attention refocused on the forthcoming close encounters due successively in early March when watchers north and south of the equator would have equal opportunity to view the comet via their domestic TV screens—assuming the key probes continued to function.

Meanwhile, NASA's old Pioneer-Venus probe was busy observing the comet extraterrestrially from its Venusian orbit and transmitting data back to Earth about the rate of erosion on the head of the comet by the solar wind. On 19 February it was estimated that the comet was losing 30 million tonnes of water per day, but, surprisingly, this figure doubled during the next few days.

As February gave way to March, there was a new sense of expectancy of great things to follow. Due to advanced TV publicity, this was in part shared by the lay public who, while still complaining about earlier difficulties of standing outdoors in the chill night air trying to catch a rare glimpse of that elusive body barely visible in binoculars *if one could find it*, were quite prepared to chance the experience via their TV screens and share vicariously the potentially unique event with astronomers at mission control.

Hoyle and Wickramasinghe had just published their new book *Living Comets*, and in the press and on TV there had been speculation whether comets were likely cosmic nurseries, plague banks—*or both*. Of the five probes approaching the comet, only the two Russian Vegas and the European Giotto would get close enough to resolve details about the comet's nucleus. By implication, if these succeeded, it might be mankind's first glimpse of the seat of his own cosmic genesis. The public's appetite for TV space spectaculars had been rekindled on 24 January by the breathtaking pictures of Uranus and her satellite system relayed back to Earth by Voyager-2 after its amazing $8\frac{1}{2}$-year space voyage. Then again four days later this interest had been reinforced by the Challenger shuttle disaster to which they were vicarious eye-witnesses shortly after lift-off.

The first major encounter was by Vega-1 which passed the comet within 8,890 km on 6 March. Although the news and TV media flashed round the world some dramatic false-colour pictures representing the densities of various gas envelopes immediately surrounding the comet, disappointingly there was little evidence forthcoming about the physical nature of the dust-hidden nucleus.

Because the nucleus appeared to be enveloped in a thick cloud of dust, there had been difficulties in estimating its size and shape. A further report confirmed that the nucleus was probably a single discrete object, and preliminary measures by the Soviet controllers seemed largely to bear out the earlier British radar observations that it was definitely an elongated mass, measuring about 15 km along its major axis and perhaps $3\frac{1}{2}$ to 4 km along its minor.

On 8 March the Japanese Suisei craft passed the comet at a distance of 151,000 km and mapped out the comet's tenuous hydrogen cloud to a diameter of nearly 20 million km. Early results revealed the cloud to be alternately expanding and contracting, and this was taken as an indication that the nucleus was a rotating body and confirmed a fact already anticipated. Further observations via Suisei indicated that the comet only emitted gases when an active side of it faced the Sun and the period of its regular "breathing", timed to be about $52\frac{1}{2}$ hours, might be a true reading of the comet's actual rotation period. This period was much longer than some had fixed it from terrestrial-based observations on the behaviour of jets emerging from the comet's head.

An unexpected feature of Suisei's encounter were hits on the spacecraft by at least two cometary dust particles, each measuring 1 to 2 mm in diameter. Even these small particles had sufficient impact to tip the spacecraft nearly 1° out of alignment.

A day later, on 9 March, Vega-2 was manoeuvred to pass the comet at a distance of 8,030 km. This was a disappointingly large stand-off distance. With one successful encounter to their credit, it was hoped that the Russian mission controllers might now

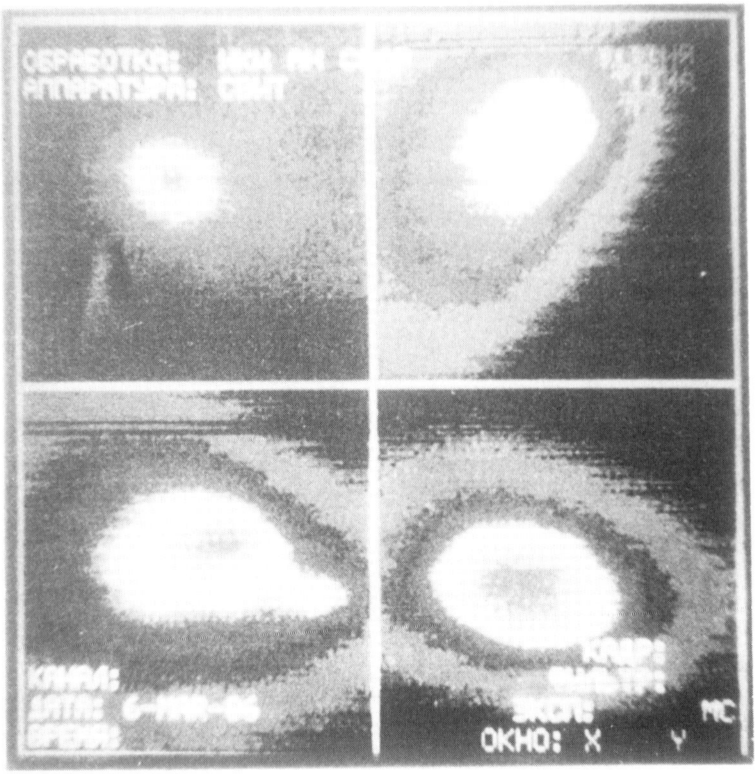

134 Four TV-screen images of Halley's comet relayed by Vega-1 on 6 March 1986.

135 Layout of Vega-1 and -2. Insets *a, b,* and *c* illustrate the spacecraft's orientation relative to the nucleus; the x, y, and z axes; and the velocity vector (v).

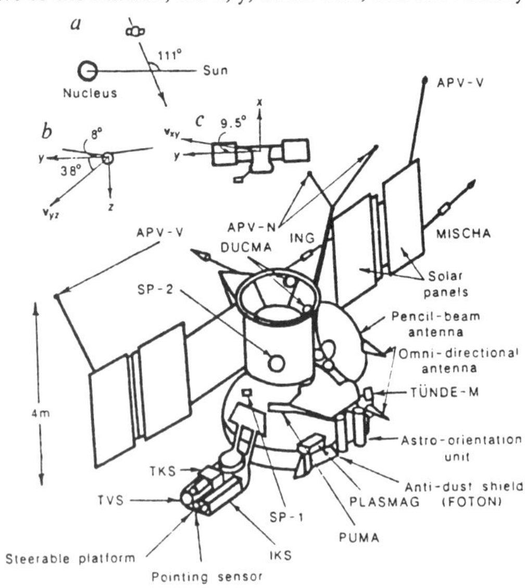

have been tempted to risk all and chance a pre-Giotto "Kamikaze" mission at much closer range.

During final approach there were some technical problems with transmission of pictures, but 7 minutes from closest approach the problems were overcome. In the event, even at the distance of 8,030 km, Vega-2 suffered heavier damage than anticipated from the comet's outer dust cloud, and this was an ill omen for the Giotto mission four days later. Power from Vega-2's solar panels was reduced by about 40 per cent. Two experiments were disabled, and for a brief period the Soviet controllers lost all communication with their craft. Nevertheless, Vega-2 sent back 700 pictures and, in spite of some earlier reports of a twin nucleus which later proved erroneous, finally confirmed the nucleus was indeed a single, solid body.

Another early report suggested the comet's real nucleus, under all that thick dust, might be smaller than first announced, perhaps only 3-4 km across the major axis. Later dimensions for the comet's nucleus were given out as 12 km × 7 km. Clearly there were ambiguities, and it was apparent that the Giotto encounter was going to be a deciding factor in deciphering the size and true structural nature of the inner part of the comet.

Meanwhile, on 11 March, Suisei's sister-craft Sakigake made the second Japanese encounter through the outer part of the comet at a distance of 7 million km and continued the programme of mapping out the characteristics of the tenuous gas cloud. At the same time NASA's International Comet Explorer, which in 1985 had made the successful encounter with comet Giacobini-Zinner, was passing between Halley's comet and the Sun, recording the speed and strength of the solar wind impinging on the comet's head. When later collated, the results of the various international probes would provide a dossier of evidence that would take months to analyze. It made the Halley rendezvous one of the most successful co-operative ventures ever undertaken.

Hard on the heels of Vega-1, and bringing things down to Earth again, was a press-inspired comedy of errors which had echoes of the Miners' comet back in 1910 (note 30). It occurred because of the total absence of up-to-the-minute pictures of Halley's comet that might be termed "dramatic". Indeed, this had been a world-wide problem facing the media since 1985 when editors, anxious to implement the old axiom that one good photograph is better than a thousand words, badgered in vain for one.

Through November-December 1985 and then into January-February 1986 the situation had not much improved. Editors and public alike were totally unimpressed by the only available current photographs of Halley's comet which suggested it was some kind of small illuminated tadpole. Comets, it was pointed out, were supposed to have long, fiery tails! When *Time* magazine ran a special feature on the comet on 16 December, its picture editor was equally frustrated by the lack of a dramatic photograph to dress up the magazine's famous front cover. Instead a staff artist was commissioned to create one. The result was a representation of the comet wrapping itself round the Earth like no other comet ever seen—except in the febrile imagination of William Whiston circa 1695.

This rendering on *Time*'s cover must have been intended as a little tongue-in-cheek, and no one took exception to the artistic licence exercised—bearing in mind another old axiom that all good art is a little cheating. To beef up the hard factual story inside, *Time* resorted to dusting off an old picture of Halley's comet seen in full bloom in May 1910 as every other publication was doing.

By early March 1986, picture editors were getting a little impatient with astronomers—especially after the new post-perihelion wave of optimism in mid-February when the re-emerging comet was reported with seven tails. At this time were conjured up instant visions of great multi-tailed comets like those of the past illustrated in old books. Alas, in actuality the comet was still only a seven-tailed tadpole.

Even the false-colour photographs taken by Vega-1, dramatic enough for astronomers, were a disappointing anticlimax to the press agencies. Prime source of this trouble was that Halley's comet in the Vega pictures *did not look anything like a traditional comet.*

Now the mischief began. To offset this identity problem someone at the Associated Press had the bright idea of juxtaposing the Vega-1 picture with something looking a little more traditional—*in fact an old faithful of Halley's comet in 1910 suitably colour-doctored.* This was done without specifically mentioning the fact in the accompanying caption. As a consequence, when the juxtaposed montage was flashed round the world, its subscribers accepted the 1910 picture as a genuine long-shot Vega-1 rendering (Fig 137).

From Britain to Australia the press marvelled at the photograph. Excited self-publicists on TV and on radio fell into the same trap and had a field-day explaining the wonders of modern technology: how it was now possible to send such a dramatic picture of the comet and its long spectacular tail, *a feature quite invisible from Earth*, across such a great distance in space.

136 In December 1985, when Halley's comet was still a faint smudge against the background stars, it was afforded the ultimate accolade as a transitory celebrity when it appeared, with considerable artistic licence, on the front cover of *Time* magazine.

Not until more sceptical, more experienced, comet astronomers had chance to get a word in, did the bubble burst. The Associated Press, when challenged, had to admit ruefully they had blundered. In a correction story they explained that in their anxiety to service clients with maximum speed, an agency photographer had taken the picture off a TV screen and had failed to check its authenticity. It later emerged that the photograph circa 1910, enhanced and doctored with brilliant false colours, was reproduced from a current comet book using the same dodge. When the story got out, the public's faith in the media and in Halley's comet took another battering. (*see* also note 31)

137 A newspaper cutting dated 7 March 1986 carrying the juxtaposed pictures of Halley's comet circa 1910 above an image obtained via the Vega-1 probe (*see* text).

Friday March 7 1986 25p

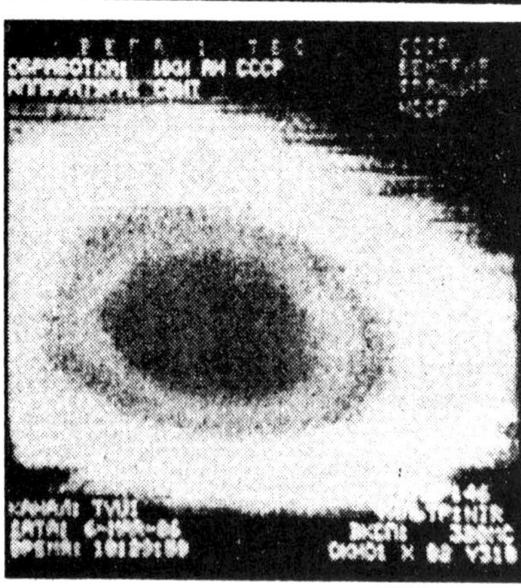

Core of the comet

PICTURES of Halley's Comet (above) taken yesterday by the Soviet space probe Vega-1 show that it has a double nucleus (left), said a Russian scientist.

Mr Lev Mukhin, a member of the Soviet team which sent up the probe in a project involving eight other European countries said that previously scientists had known only that the nucleus was capable of disintegrating.

The nucleus appeared on monitoring screens at the Moscow Institute of Space Studies as brilliant red, surrounded by a flare of yellow enveloped in green. Vega-1 took the first colour pictures of the interior when it was 4.4 million miles from the comet.

The space probe was 93.8 million miles from earth at the time

Astronomers, case-hardened to such press blunders, shrugged off the episode with a wry grin; scientific faux pas are to be expected as part of the normal order of things in most newspapers! To comet astronomers, the impressionistic false-colour pictures from Vega-1 accompanying the sleight proved more fascinating than any priceless Monet hanging in the Louvre.

As 13 March 1986 dawned, tense mission controllers in the Headquarters of the European Space Agency in Darmstadt, West Germany, were preparing Giotto for its final approach, timed for midnight GMT. Gathered round them were TV crews ready to transmit a few hours from now to at least 50 nations whose combined viewing audiences were expected to number more than a thousand million. The event, already billed as "Drama from Darmstadt", was expected to provide one of the great moments in the history of space encounters.

Giotto's planned encounter at 500 km was the result of compromise to suit the various needs of the experimental groups from West Germany, Switzerland, Britain, France and Ireland. Experimenters interested in the comet's dust content would have preferred to have gone in much closer, but the primary restraining influence was the safety of the onboard camera which if hit by a single stray dust particle might totally wreck the fundamental aspects of the mission.

In view of the hits and damage incurred by the Japanese Suisei and the Vegas, the open question was whether Giotto could survive the expected heavy bombardment from the comet's dust. Since Vega-2, the Russian spokesmen, at least, were quite sure that Giotto was now a doomed Kamikazi mission.

The gathering press as usual could be relied upon to contribute their fair share of gaffes. One journalist, totally misled by the well-publicized remarks of the Russian commentators whose remarks were translated so as to describe Giotto's mission as "suicidal", took the word much too literally and got it into his head that Giotto was a *manned* probe. He queried, in view of the recent Challenger disaster, whether men should really be allowed to risk their lives chasing after comets...

Giotto was a complex instrument package comprising 11 different experiments. These ranged from the prime optical mission to obtain direct images of the nucleus to others involving spectrometry, particle and plasma analyses and magnetic studies. All their ultimate success relied on the correct functioning of the spacecraft's transmission system, which was its most vulnerable link. If Giotto wandered off course, or was nudged off course by a particle of dust, by as little as 1° (as Suisei had been), then all communication back to Earth would be interrupted. Since the gathered data was not to be stored on board but transmitted directly, this would be irretrievably lost—even if at a later time the craft restabilized itself.

The run-up during the previous week had not been without alarms for the mission controllers. On one occasion, for no apparent reason, the spacecraft suddenly switched itself off and the signal lost. Engineers beavered away for half a day to try to correct the fault before Giotto was finally coaxed back to transmitting again. Clearly then, on the night and for good reason, there were to be fingers crossed at Darmstadt...

At first, the final approach went according to all expectations. After Giotto passed through the comet's bow shock wave (caused by the solar wind interacting on the cometary gases) it entered in turn through the different zones of the outer more tenuous parts of the comet, already explored and mapped out a few days previously by the two Soviet Vega probes.

The speed of the solar wind was measured at 380 km per second, but closer to the comet its velocity was reduced to 350 km per second owing to the more dense cometary particles encountered there.

The strength of the anticipated bow shock wave proved less than forecast, and, if anything, the terminology was in error—the phenomenon being better described less dramatically as a "bow wave". Another surprise was that Giotto encountered the bow wave at a distance of 1.15 million km out from the nucleus. This was *twice* as far out as had earlier been mapped by the Vegas. It could be explained by the fact that at the time of both Vega encounters the solar wind had been stronger, and this had the effect of confining the comet's gases more inside the coma.

Meanwhile, as the probe moved closer, measurements revealed a distinct slowing down of the velocity of the solar wind till it was reduced to 260 km per second.

138 An artist's impression of a controller and his monitoring screens during Giotto's historic encounter with Halley's comet on 13 March 1986.

139 The layout of the Giotto spacecraft.

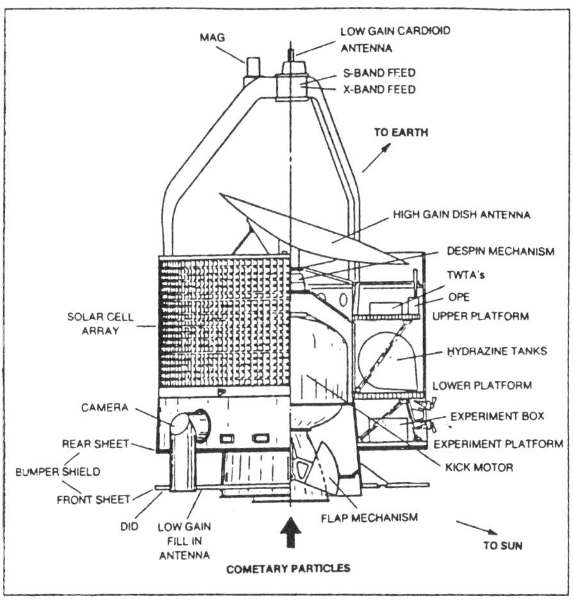

It had been forecast that at a range of 10,000 km from the nucleus there should be a well-defined transition in the make-up of the comet's gases—changing from intermixed ionized gases (originating from the interaction of the solar wind) to largely neutral gases surrounding the nucleus. This change, however, was not registered till 65 seconds before closest encounter, at a target distance of 4,700 km, when at the same time the magnetic-field strength faded to zero.

But what of the comet's dust particles? It was the dust which could make or break the success of the encounter...

The first dust encounters were expected at a range of 500,000 km—or time-wise about 2 hours before target pass. Both Vega craft had suffered impact damage to instruments not specifically protected. By comparison, Giotto was an armoured probe. To reduce the risk of malfunction, its instrument package was shielded with some very sophisticated bumper armour designed to soak up the dust impacts.

Giotto had in fact a two-layer protection system. First line of defence was a thin aluminium sheet, then, 23 cm behind, a 15-cm thick Kevlar epoxy and polyurethane foam shield. In principle, dust particles should vaporize as they passed through the aluminium, and this resulting vapour would then be dispersed by the second layer. How well in practice the system would stand up to bombardment depended on the density and size of the cometary particles encountered.

It was not until Giotto closed to 290,000 km (nearly half the originally estimated distance for impacts) that it recorded and absorbed its first dust particle but without any dire effects. The protective system seemed to be working as expected.

During the next half hour, as mission controllers exchanged tense glances, 14 more hits were registered and absorbed. But this count was far fewer than earlier forecasts—perhaps only one-tenth as many. Clearly, the structure of Halley's comet seemed to be different from its theoretical model.

Were comets a lot less dusty then than had been imagined?

If there was now a lingering doubt, a sure reminder that comets contained *some* significant dust particles came when Giotto, closing in at almost 245,000 km per hour, passed the

8,000 km mark and the first sizeable particle impacted and penetrated the main bumper. Fortunately this was checked by the secondary Kevlar shield.

It was estimated this particle weighed 30 microgrammes. It proved to be only the advance guard of an onslaught of larger particles that peppered Giotto immediately after.

Giotto, it was plain, was now in the thick of it and anything could happen...

After a hundred or so particles, travelling at speeds 50 times faster than terrestrial bullets, had penetrated the defences of the outer bumper, a larger than average one impacted and dramatically sent Giotto reeling, causing its signals to veer away from the direction of the Earth.

This happened at exactly 3.1 minutes after GMT midnight and just 2 seconds before closest approach to the nucleus. The screens monitoring the comet momentarily flickered with the dying picture; then it went blank. One thousand million viewers all asked themselves the same question. Had Giotto on its Kamikaze mission amid the comet's lethal dust perished for good?

This was the nightmare moment long dreaded by mission controllers, yet perhaps secretly welcomed by TV producers. Here indeed was the sure ingredient for a nail-biting cliff-hanger to maintain audience suspense. Space spectaculars that ran too smoothly lacked human interest and were known to induce yawns and a massive channel switch-over.

When the screens monitoring pictures of the comet finally went blank, it was followed soon after by a voice-over confirming laconically that all telemetry with Giotto was lost. In the confusion, this message was unfortunately missed by many TV stations whose link-men were therefore unsure about precisely what had happened.

Also missed was a second message to the effect that Giotto's signals were intermittently coming through again. These, however, as Giotto yawed along its track, were too weak to transmit pictures.

There was to be no encore when Giotto re-emerged intact from the coma region, still spattering its signals from the far side of the comet.

Then, 32 minutes after first loss of signals, Giotto's damping system finally stabilized it, and full control was restored; but still no pictures. As far as the world-wide TV audience was concerned, the show was over.

Giotto *had* survived but only just. At least five groups of instruments were destroyed while others still operating were damaged. In spite of the last-minute setback, Giotto had performed superbly and delivered scientifically all expected of it. Right up to the final two-second count-down, Giotto's pictures, totalling more than 2,000 images of the comet plus a torrent of supporting data, had kept coming in. As the mission progressed, the comet pictures transmitted at intervals of four seconds had loomed larger and larger till they yielded the best ever views of the nucleus. By the time the screens went blank, the mission was already an unqualified success story.

Final results, nevertheless, were still several weeks away. In some instances it would take months to analyze all the data and correlate it with data recorded by the Soviet, Japanese and US probes.

Giotto's ability to close into its target was greatly assisted by the exchange of information between the ESA mission controllers and the Soviet Space Agency, Intercosmos, who generously passed on all the relevant data from their Vega encounters. Vega-1, directed from ground-based observations, had ended up some 480 km closer to the sunside than its targeters intended it, and this data plus that from Vega-2 allowed the ESA navigators to up-date Giotto's path and ensure a really accurate encounter.

While the main operation had run comparably smoothly, it had not been without some little-publicized and nail-biting incidents in the days preceding.

One unsettling incident had occurred at 2300 hrs GMT on 11 March when, with some of the spacecraft's 11 experiments turned on and already transmitting, mission controllers suddenly lost all contact with Giotto. The cause of this hiatus was later traced to the intervention of a rogue bulldozer which had dug up a key telephone cable in Australia. This cable was the very link between the European Operations Centre at Darmstadt and the relay station in Australia receiving Giotto's signals from across 150,000 million km in space.

140 Halley's comet photographed from Western Australia on 13 March 1986, the same day Giotto made its historic encounter.

To compound difficulties, both separate receiving antennae in Australia were linked for a short distance by the *same* stretch of cable. By the intervention of Murphy's well-known law, this particular section was the very one damaged by the bulldozer!

A contingency plan to switch communications links, using NASA's long-established deep-space network, had been arranged earlier, but even this back-up system failed at first to pass on the Giotto data. For a nerve-wracking half hour the European technicians battled to sort out the various computer incompatibilities before the problem was overcome and Giotto's signals came in again loud and clear.

Although space technologists and comet astronomers were euphoric with Giotto's success, by contrast, the lay media were ambivalent and cautiously measured in their praise. While accepting it was a major technical triumph for man to reach out from Mother Earth and peer into the heart of Halley's comet, they had reservations.

Part of their misgivings stemmed from the TV coverage. In the days following, critics and audiences alike admitted to sometimes having been greatly puzzled by events. "It was a great show if you were an *expert*," summed up a TV critic with ironic backhandedness. Indeed, a common complaint raised against the TV link-men in touch with Darmstadt was they were accused of addressing their lay audiences in an only part-comprehensible jargon—a language really only fully understood by the space-buff cognoscenti. This engendered resentment and made the lay audience feel left out, unable to share fully in the excitement. Another general misgiving was directed to what was actually presented to them on the TV screen, for it had little resemblance to the photographs of Halley's comet as seen in 1910...

Whereas the TV coverage of the Apollo missions in the late 1960s and early 70s had shown immediately recognizable Moonscapes inhabited by astronauts and the Voyager probes had revealed vivid close-ups of distant planets and their satellites, the Vega and Giotto transmissions, using their false colour imaging, made Halley's comet appear uncomet-like and bizarre.

A leading TV critic in one of the Sunday heavies disparagingly likened it to "a computer weather-map printout of the Gulf of

Mexico" *or* "possibly the cross-section of a plaice cooking in a microwave oven". The problem was he could not quite decide which. Another droll critic dismissed it as "Picasso's idea of a fried egg". The fried-egg image was also vivid to others: "... a badly fried egg nestling in a dollop of tomato ketchup", elaborated a third critic. A fourth, more generously concluded: "I couldn't understand a word of what was going on, but I didn't think the show was all that bad."

141 A view of the innermost coma and nucleus of Halley's comet as scanned by Giotto from a distance of 25,650 km. The original picture was transmitted in pseudocolour images which gave it a bizarre appearance on TV screens and led to confusion and disparagement among lay audiences and critics. For scale appreciation, note that each side of the frame represents 111 km.

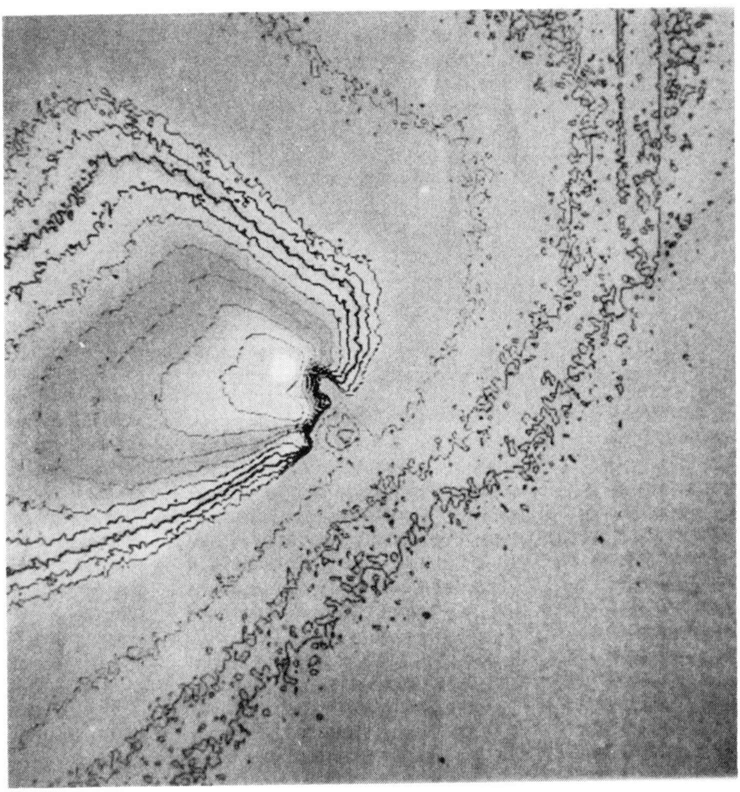

The last photograph taken of the comet's nucleus was from a distance of 1703 km and therefore not as close as originally hoped for. Even so, the images provided a wealth of resolvable detail. The nucleus itself was perfectly visible and revealed what appeared to be pit-like structures on its surface, some about one kilometer across. When an emotion-charged voice-over from Darmstadt exclaimed he could see "mountains, hills and craters" and "incredible detail", it was all heady stuff for those looking in who had spent most of a lifetime speculating about cometary structures. The old nagging query: loose sandbank or a discrete solid body (Fig 111) with the lingering question-mark still against it just days beforehand, could now be settled definitively.

A major surprise had been the very dark appearance of the comet's nucleus. Although no one had really expected the nucleus to reflect sunlight like a dazzling ice-ball, it was much darker than most had anticipated. This in turn gave rise to some spontaneous voice-overs from Darmstadt: "...velvet black!" came one view; "...blacker than black!" another; "...the darkest dark you can imagine!" chipped in a third anonymous voice. In more scientific terms its albedo (or reflecting power) was rated not greater than 5 per cent whereas the Earth itself reflects about 34 per cent of the light it receives from the Sun.

Ten days before the first Vega fly-by, Hoyle and Wickramasinghe had predicted that the nucleus of Halley's comet would appear very dark owing to the presence of organic matter. While this was now proven true, at least as far as colour was concerned, others previous to this had also predicted the same thing. Beforehand, it was generally supposed that comets, surface-wise, might be relatively dark-bodied objects, for there was a distinct possibility they were related to a class of dark meteorites known as carbonaceous chondrites. These meteorites possess some unique characteristics quite different from other meteorites. On occasion they had provided some tantalizing laboratory evidence which some claimed indicated true organic structures. Nevertheless, the orbits of these meteorites, unlike those of meteors, had never been positively linked to the orbits of comets. Some of the asteroids were also known to be very dark bodies, including the recently discovered 1983 TB. Were then some of the

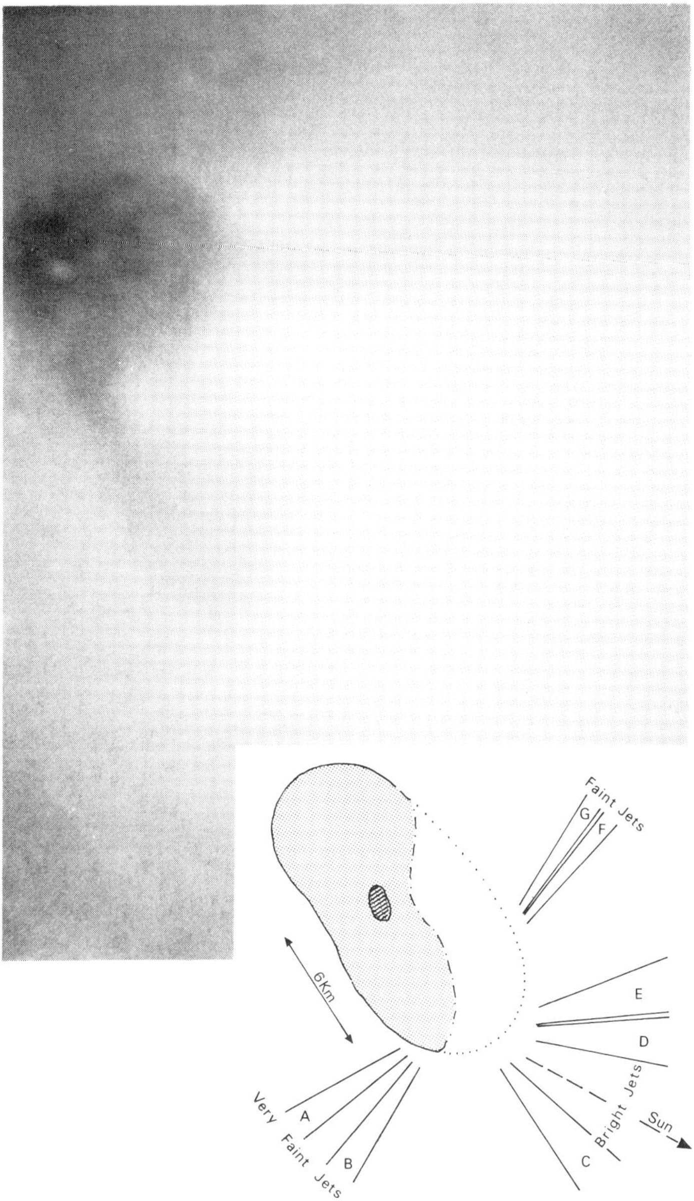

142 (upper) The dark semi-obscure night-side of the nucleus of Halley's comet as imaged by Giotto's camera from a distance of 14,000 km on 13 March 1986. Some rugged circular features are just resolvable in the hemisphere farthest from the Sun. (lower) An outline sketch depicting features interpreted from the best images and showing dust jets not properly visible in reproductions.

asteroids also related to comets and carbonaceous meteorites?

These speculations were now back for airing as Halley's comet seemed to be providing circumstantial field-evidence to link these various bodies.

When Vega-1 had made its run past the comet, the nucleus was only seen in vague outline, being mostly hidden by the comet's surrounding dust cloud. During the Vega-2 encounter, visibility was better and two jets emerging from the nucleus were recorded on photographs. It was this twin phenomenon that at first was erroneously interpreted as indicating the comet had a double nucleus.

Giotto was fortunate to encounter the comet when it was less active, and the nucleus was clearly visible as a discrete body about 15 km long by 8 km broad. Physically it gave the appearance of a very irregularly indented, dark body emitting several bright jets of gas and dust. It was described by some watching the monitors as black and potato-shaped while others at Darmstadt saw it rather as peanut- or banana-shaped.

The closest pictures revealed a wealth of distinct structural features. These appeared as ring-like depressions but are unlikely to be craters as we know them on the lunar surface. They are probably vents, or nozzles, through which the inner material is ejected when the comet is activated.

Unfortunately the closest pictures taken by Giotto did not show the nucleus completely. The closest picture showing any part of the nucleus was taken at a range of 1703 km or time-wise 12 seconds before target encounter. This, nevertheless, was good enough to resolve surface detail down to about 100 metres. Giotto actually passed the nucleus at a distance of 605 km, within 70 km of its final targeting.

With due confirmation that Halley's comet had a black surface, just as Hoyle and Wickramasinghe had forecast, their speculative theories about life from space via comets received a new boost. It was conjectured that such a dark reflecting body might be strongly indicative of a tarry organic crust overlying its surface, probably the result of repeated erosion of the comet's volatiles during its periodic visits to the inner solar system. It was also conjectured that other short-periodic comets like Encke's could be expected to have similar black-surface characteristics. It was surmised that such a black surface would absorb extra-large amounts of incidental sunlight which, if the comet were formed mostly of pure ice, would lead to its evaporation at a much greater rate than actually observed. This anomaly now provided a new puzzle for comet modellers.

During late March while the comet was under ground observation from large telescopes in the southern hemisphere, the life-from-space speculations received yet another fillip.

In Australia, the 3.9-m Anglo-Australian telescope photographed the comet in the infra-red region over three nights from 30 March. The spectrum obtained revealed that the dust cloud carried a characteristic fingerprint which emitted at a wave-length of 3.38 microns, identified with a wave-length occupied by carbon-hydrogen. Surprisingly, the same spectrum showed no peak at 3 microns, the wave-length of water-ice particles expected to show up in fair abundance *if* the comet were mostly an icy snowball as envisaged by Whipple and some nineteenth-century astronomers.

These new observations, if taken at face value, seemed to confirm a prediction made some years earlier based on laboratory studies that such a spectrum might be expected in the infra-red of comets owing to the presence of bacterial dust. Not surprisingly, Hoyle and Wickramasinghe claimed the new Australian observations now provided unambiguous evidence that organic molecules were to be found in Halley's comet.

143 Shortly after the Vega/Giotto probes, spectra obtained via the 3.9-m Anglo-Australian telescope provided some controversial new evidence about organic matter in Halley's comet and attracted further support for Hoyle's and Wickramasinghe's theories.

New evidence of organic material in Halley's comet

Sir Fred Hoyle — in from the cold

By Anthony Tucker, Science Correspondent

Observations of the departing Halley's Comet ade by the Anglo-Australian telescope have confirmed that the bulk of the material in the comet's dustcloud is organic. The "dirty snowball" theory that comets are made largely of water ice and interstellar dust has finally melted away.

Unique high resolution measurements of the comet's emission and absorption in the infra-red range have revealed that the dust cloud has a broad but strong "fingerprint" in its spectrum at a wavelength of 3.38 microns — a wavelength associated with carbon-hydrogen bonds. The spectrum has no peak near 3 microns — the wavelength of water ice particles.

The Australian observations were carried out by Dr David Allen and Dr Dayal Wickramasinghe. First confirmation of the measurements reached this country on Tuesday and Professor Chandra Wickramasinghe — Dayal's elder brother who, with Sir Fred Hoyle, has long maintained that cometary material must be composed mainly of complex organic substances — said that the observations "confirm the essentially organic nature of the material beyond any question."

The space probe Giotto found that although some water molecules appeared to be present the bulk of the material was composed of carbon, nitrogen and oxygen. Because of the elementary nature of the measurements, they are not in conflict with the Australian observations.

The Australian study compares the detailed structure of the comet's infra-red fingerprint with predictions made five years ago by Professor Wickramasinghe and Sir Fred Hoyle, from laboratory studies, of the infra-red spectrum of bacterial dust.

Professor Wickramasinghe said: "Although it seems rather bizarre, every new observation seems to give additional support to the theory that cometary material is very complex, certainly organic, and possessing features which are very like those we might expect from the residues of living material, such as bacteria and viruses." Since he and Sir Fred Hoyle correctly predicted the very black nature of the cometary nucleus, the probable presence of jets of material, the absence of water and the dominance of complex organic substances—predictions later confirmed by Giotto and the Russian Vega space probe as well as by the latest measurements—they may now be allowed to come in from the scientific cold.

However, space experts in Britain remained sceptical yesterday about the bacterial hypothesis. Dr John Davis of Birmingham University's department of space research said that while comets do contain organic material and the general view now is that black surface of Halley's core consists of an organic complex of some kind, there is no evidence yet that this material is bacterial or related to viruses.

"I don't think many people would support such a theory on the present evidence," he said.

Opponents to the life-from-space ideas via comets immediately took issue. They, in turn, claimed these particular spectra proved nothing, and it was still not as clear cut as Hoyle and Wickramasinghe generally supposed. These opponents pointed out that astronomical infra-red spectra were subject to great uncertainties in interpretation. They emphasized that dark organic molecules might by synthesized inside a comet (perhaps by the same mechanism as was supposed to be occurring in interstellar clouds) and, therefore, biological processes are not necessary to explain such observations. They stressed that the evocative word "organic" provided an annoying misnomer in the argument, suggesting it always implies a ***biological*** origin when really only "straightforward chemistry" is required.

144 During March 1986 Halley's comet brightened to 3^m and developed a tail 10 to 12° in length. Although it provided some impressive enlarged pictures via bigger telescopes, this photograph taken with a small camera on 23 March 1986 from Virginia USA shows the comet more realistically as it appeared in small binoculars or to the naked eye. The quadrangle of four brighter field stars to the comet's upper right are Zeta, Tau, Sigma and Phi Sagittarius (*see* also Fig 145).

145 Halley's comet photographed from Mt Stromlo, Australia, on 21 March 1986 showing the comet with a 10° tail and the star clouds of the Milky Way above. Note the quadrangle of field stars immediately to the upper right of the comet also seen in Fig 144.

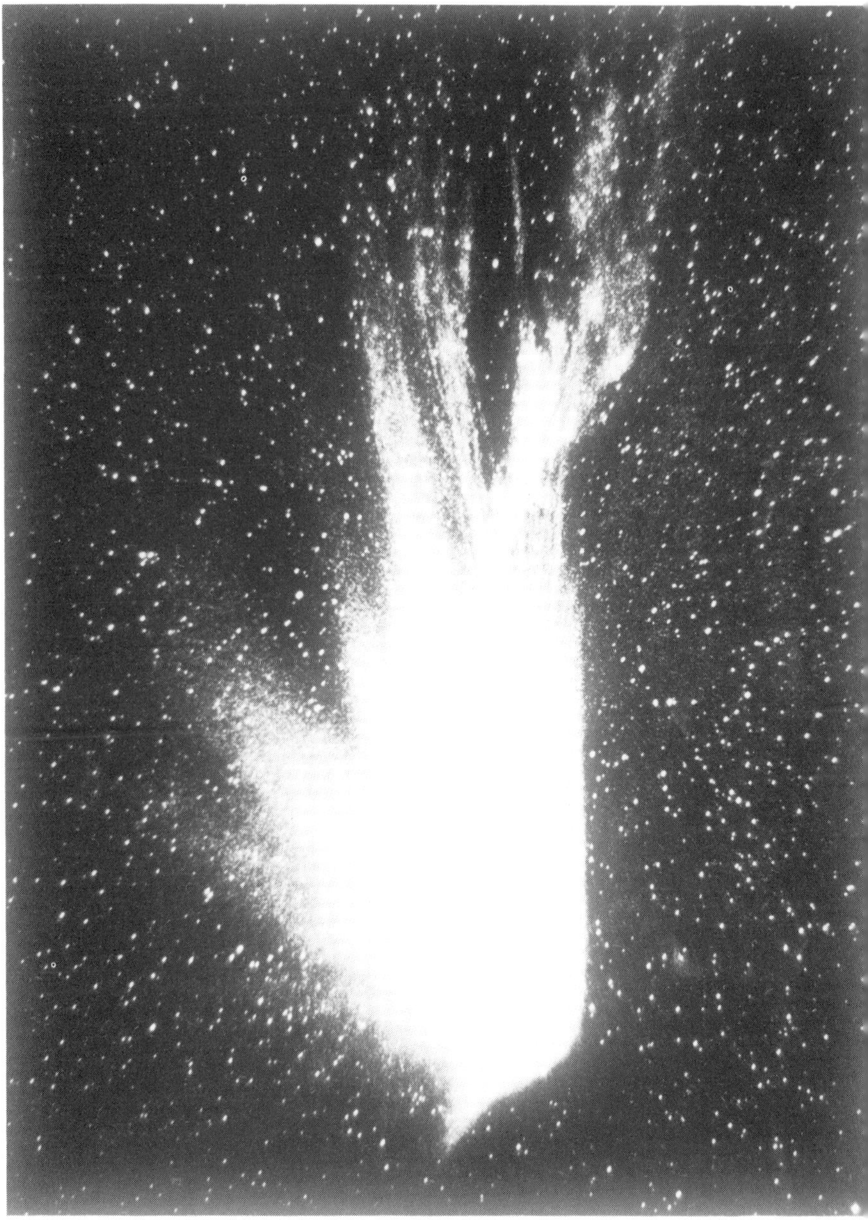

146 When Halley's comet emerged in the morning sky after perihelion, its tail had developed into a broad multi-pointed fan with one spike extended towards the Sun. This enlarged photograph taken on 22 February 1986 by the 1.2-m Schmidt telescope at Coonabarabran, Australia, showing disruptions in the gas tail, is probably the most dramatic taken during the entire apparition.

Although fairly complete summaries of the various experiments carried out by the different spacecraft were readily available for immediate assessment, they still required careful analysis in the days and weeks following. Two months later an update of all these experiments was written up and presented to the scientific world in a series of 38 comprehensive papers via the premier journal *Nature* in its issue dated 15 May 1986 (Vol 321).

While even some of these findings were still relatively tentative, the package provided a tour de force of information of magnificent proportions and underlined again the unqualified success both in terms of scientific data and co-operative effort at international level.

It was now made clear that Giotto's camera had failed some 12 seconds before closest encounter owing to a glitch in the spacecraft's power-system and its pictures only represented one side of the comet's nucleus.

The Vega pictures, the best taken from 8,000-9,000 km, did not resolve nuclear features in as much detail as Giotto's, but it was anticipated that after additional processing they would provide a fully comprehensive coverage of all surfaces. The Vegas had indicated the nucleus to be a body 14 km long × 7.5 km wide × 7.5 km thick, with a probable accuracy in any dimensions of ±1 km. The Giotto findings largely confirmed this, indicating a length of 15km, but unfortunately the width and thickness were still uncertain.

When Vega-1 encountered the comet on 6 March, it met the dust coma at a distance of about 280,000 km from the nucleus. At this time the comet was very active and emitting dust at the rate of 10 tonnes per second. Three days later, when Vega-2 encountered the comet, it was much less active, and the measured gas and dust were only half that encountered by Vega-1.

Vega-2, using its special infra-red equipment, measured the temperature of the nucleus to be 300-400 K* whereas, previous to this, theoretical models for an ice-ball nucleus predicted a lower figure of 180-200 K. Clearly the very dark obscuring mantle had some effect on temperature.

*To convert K into Centigrade, subtract 273°

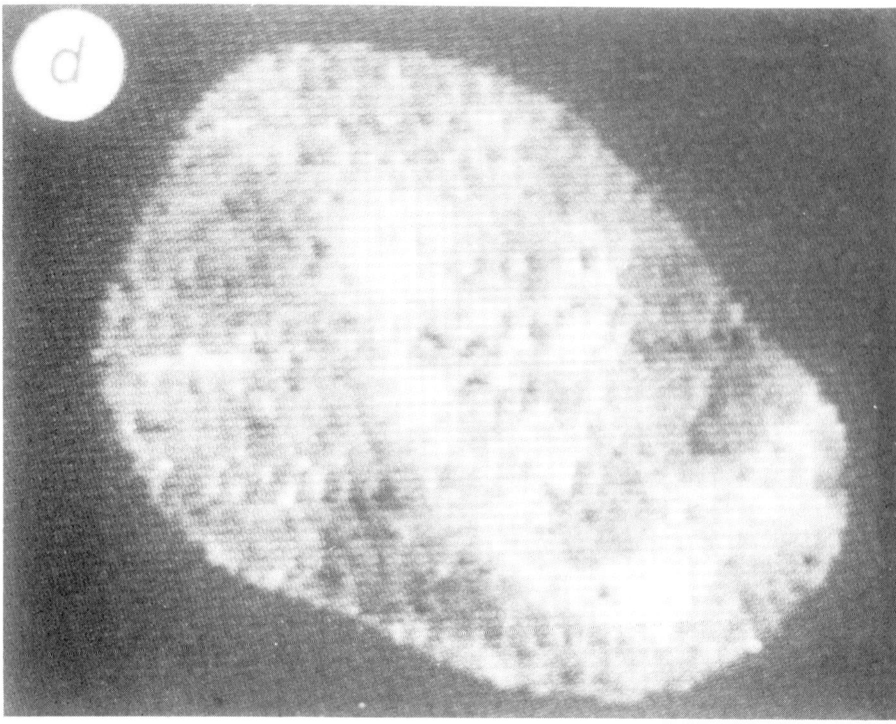

147 The nucleus of Halley's comet as it appears in processed images obtained from Vega-1 on 6 March 1986.

The Vegas also confirmed a rotation period of about 53 hrs ±3 hrs (again confirmed by the Giotto figure of 52 hrs) and an albedo round 4 per cent—about the same kind of low reflectivity possessed by the thin, faint rings found round the planet Uranus.

Vega-2 obtained 3,600 spectra in the near ultra-violet and visible region. These showed the most prominent emissions were OH, NH, CN, C_3, CH, C_2 and NH_2, previously recorded in other comets. Vega-1 returned data for about 1,000 dust particles, but there were fewer counted by Vega-2 owing to voltage problems with the spacecraft. Their respective dust analyses detected the presence of C, O, Na, Mg, Si, Ca, and Fe. Giotto's analysis of the dust particles showed most of these to be rich in light elements such as H, C, N, and O.

In the coma, H_2O was confirmed to be the dominant parent molecule, and from all the data it was clear that Halley's comet appeared to be a predominantly water-ice body—a semi-pristine, 4.6-thousand-million-year left-over from the begin-

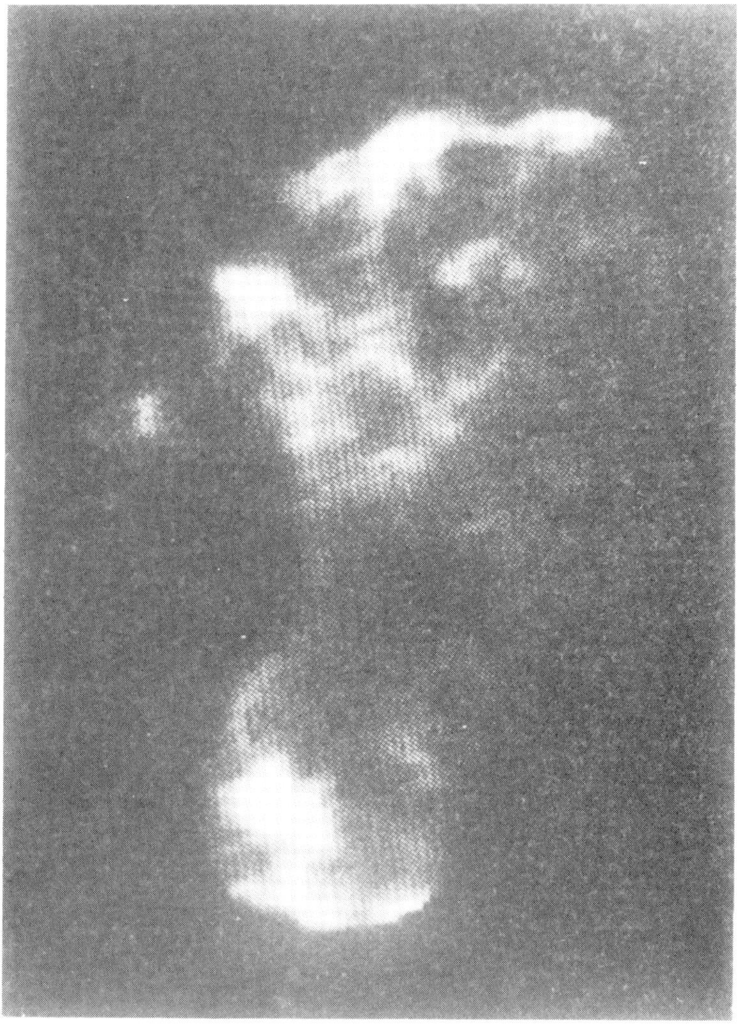

148 A tantalizing glimpse of the rugged nucleus of Halley's comet derived from the processed images provided by Vega-2 on 9 March 1986.

nings of the solar system. While perhaps 80 per cent of the comet was water-ice, there was a strong intermixture of solid particles, and the nucleus seemed to be a much dirtier snowball than envisaged by the Whipple model. The nucleus was overlaid with a dust crust which gave it its low albedo. How thick this crust is and whether it is evenly distributed still remains uncertain. It could be as thin as 1cm or as thick as 1m. These questions, among scores of others raised, now nagged to be answered by future, more advanced, probes to comets.

Surprisingly, Halley's comet showed little or no sign of sodium which is often markedly present in bright new comets near time of perihelion (*see* note 31).

Of great significance was confirmation by the Vegas and Giotto that the cometary dust *included organic material.* However, these dust particles conform only roughly to type C1 carbonaceous chondrites, but a more detailed analysis was still awaited.

In total, Giotto recorded 12,000 hits by dust particles, the first of which was encountered at a distance of 290,000 km from the nucleus. The largest particle weighed 40 mg and hit Giotto 44 seconds before closest approach. A surprise was the dust particles were generally larger than expected. Closest approach was now confirmed to be 605 km $\pm$8 km which occurred at 03 mins 02 secs GMT 14 March. The last sequence of pictures was exposed at a target distance of 1,703 km. In all, Giotto identified seven jets emerging from the nucleus on the sunlit side plus a wealth of rugged surface detail earlier vividly described by the Darmstadt controllers on the night of the encounter. However, all the pictures were still undergoing further processing, and more detail was expected to show up when this was completed.

Although the Vegas and Giotto suffered damage or deterioration, they could be recycled for further encounters. Giotto on its present path is scheduled to return to the neighbourhood of the Earth on 2 July 1990. Then, if required, it could be retargeted, using the Earth to gravity assist it, to rendezvous on 14 July 1992 with a short-periodic comet known as P/Grigg-Skjellerup, first discovered by an amateur in 1902, and whose orbit carries it round the Sun in 4.9 years. The Russian controllers were also studying future targets for their Vegas, possibly asteroids plus other encounters through meteor streams known to be associated with comets.

Another result of the Vega and Giotto encounters was the ability to fix the orbit of Halley's comet with more precision than any other of the 1,000 or so recorded comets. Until the fly-bys, Halley's comet could not be fixed in space from Earth observatories closer than 1,500 km. At the time of the Giotto fly-by, the uncertainty was reduced to 40 km. However, as time passes, the

149 A photograph of Halley's comet in January 1986 providing a view of it highly reminiscent of Barnard's picture circa 4 May 1910 (Fig 99).

uncertainty will increase again owing to perturbing influences of the planets and the non-gravitational forces induced by the force of the emerging jets. Even so, Halley's orbit will remain the best described of all comet orbits for some time to come.

Comet astronomy after the Giotto/Vega fly-bys could never be the same again. The veil of secrecy long surrounding comets had been partially if not fully raised. Three hundred years earlier, Newton's *Principia* had shown how comets moved in their paths but had not explained anything about their physical properties. Since then it had taken three centuries of intellectual endeavour before man was able to peer into the heart of a comet. In the years ahead, the Giotto/Vega fly-bys would represent the great watershed in the understanding of comets. References to "Before Giotto" and "After Giotto" would need no further elucidation in the nomenclature of the science.

The successful fly-bys of Halley's comet, plus the circumstantial spectral evidence, still did not *prove* that terrestrial life originated via the comets or that the observed dark material was actually bacterial. Proof or discredit of the idea can only be forthcoming through further space probes to exploit and extend the findings of the first fly-bys. Probes to asteroids, already scheduled, will help resolve the question whether *some* of these bodies might be related to comets and whether asteroids represent a variety of objects with differing origins.

Probes to other comets with direct orbits like Encke's comet are now a priority. On such a probe the spacecraft can be made to rendezvous with the comet at a great distance from the Sun and then accompany it on its journey inwards and then out again, keeping exact pace so as to study its behaviour throughout one orbital revolution. Other probes will be required to rendezvous with the large, brilliant comets whose periods are measured in some thousands of years to see whether these more pristine bodies, whose orbits stretch into space almost halfway to the nearest stars, differ physically from the near-at-hand short-periodic comets which visit the Sun regularly.

The ultimate probes will be to make soft landings on comet nuclei to sample the various layers. Only when this programme has been successfully accomplished will we know for sure whether comets are the source of genesis for all life on Earth.

The *Principia* was the fountain-head that set man firmly on his path to journey out from his own planet almost three centuries later, and had Newton not written it, we can be sure that the whole history of modern science and much of our civilization would have been different. One of the fascinations of studying science is looking back to examine the circumstances of revolutionary advances and noting how these often occurred at the whim of the Fates.

Newton's *Principia*, especially, has an air of serendipity to its creation. If Halley had not been beset with solving the mathematical problem, set by Wren, which sent him to Cambridge to seek advice from Newton and bestir the sleeping giant from his reclusion, *Principia* might never have seen the light of day, and the story of Halley's comet would have taken another course. Yet even before this, and what is often overlooked, the Fately sisters Clotho, Lachesis and Atropos had surely laid their plan through lesser but worthy broadminded men. If the lowly sub-sizerships at the University of Cambridge had not existed to enable the importune student to live by the performance of some menial domestic duties, while pursuing his studies, Newton might forever have remained a Lincolnshire rustic.

It was the French Encyclopaedist Denis Diderot (1713-84) who defined science as a large area strewn with dark and illuminated places. Man's intellectual labour, he believed, should have its aims, each according to talent, either to extend the limits of the lighted places or add to the number of centres of illumination. The latter was the task of the true genius.

Of the two principals in our story, Newton was the illuminator of new regions while Halley the extender of them. As scientists they provide us with sharp contrasts. Newton was the introvert—a seventeenth-century super Archimedes; Halley the extrovert—a prototype nineteenth-century man of action. They shared with their great predecessor Kepler a tendency for *incuria* and they were fallible. Newton's *Chronology of Ancient Kingdoms Amended* (1728) has absolutely no worth to it, and we can barely recognize it as coming from the same hand that earlier created the *Principia*. Halley was often no great shakes as a comet computer. He was wrong about the 1680-1 comet; he was wrong in linking the comets seen in 1264 and 1556; he was wrong yet again in linking the comets of 1532 and 1661. Had not his prediction for the comets of 1682, 1607 and 1531 proved correct, we should probably now be remembering him as Newton's sometime shadowy amanuensis who didn't quite make the grade.

Thus the lesson to be learnt is that even the greatest minds are sometimes culpable. In accepting all new ideas, the working scientist *and* the general reader need to maintain a sceptical approach and to be ever mindful of Finagle's mischievous, but wholly apposite, Laws of Science: Science is truth: don't be misled by facts / No matter what result is anticipated, there is always someone willing to fake it / No matter what occurs, there is always someone who believes it happened according to his pet theory...

Another fascination of science is that no one can predict the next great break-through or its rate of progress. H G Wells, recognized as the grand visionary of his day, was, in hindsight, badly amiss in his time-scale. No one at the turn of the century could imagine how two world wars would revolutionize technology and accelerate scientific progress beyond anyone's wildest prophecies. Space-travel itself had long been envisaged; it only awaited supporting technologies, in particular the advent of World War II to make it a reality. When Crommelin died in 1939, he could not possibly have foreseen space travel would happen so quickly, helped by unpredicted inventions such as the transistor and the still almost magical micro-chip. Least of all, he could not have imagined that some of his younger colleagues, still alive 47 years on, would be clustered round a TV monitor gazing at the innards of the comet whose anomalous motions had sorely puzzled him.

It needs no crystal ball to predict that trips to comets will soon become routine ventures and that mankind will colonize them as unique travelling habitats. So far outer space has remained free of military imperialism, but for how long? When Halley's comet swings round the Sun again in 2061, the world will have witnessed unpredictable changes. The burning question in the age of nuclear uncertainty is whether homo sapiens will still be extant to greet it. Unless by remote chance, in the interim, we are visited by some Wellsian-type comet of the circa 1906 science-fiction variety, which envelops us with its tail and exerts its benign influence over the whole mankind, the easiest prediction to make for 2061 is the immutability of human nature.

Supplementary Notes

Note 1 p. 1

The pronunciation of Halley's name has developed into one of those tiresome debates which used to involve the names of Raleigh and Shakespeare. There is evidence to suppose that some of the seventeenth- and eighteenth-century Halleys pronounced their name "Hawley"; however, contemporary Halleys seem to favour "Hali" or "Haili". In Pepys's naval minutes the name is written Hawley, but, in tune with the inexact spelling of the age, one may find in other records alternative renderings including: Hailey, Hawly, Haley, Haly and Hally.

Note 2 p. 23

Thomas Hearne (1678-1735), antiquarian and prolific author, held the position as second keeper at the Bodleian Library. He became well acquainted with Halley after Halley's appointment as Savilian Professor of Geometry at Oxford in 1704, but they were never friends. Hearne was a Jacobite and by his own admission was annoyed by Halley's frequent condemnation of Mary Queen of Scots. Because of his Jacobite sympathies, Hearne was forced to resign his position in 1716. In Hearne's often maliciously inspired *Remarks and Collections* (published Oxford 1884-1921) there are several brief references to Halley which show the pair were never on good terms.

Note 3 p. 39

What may explain past confusions about Mary Tooke's family background is that, as far as I can determine, her *grandfather,* one John Tooke, once held the title of "auditor". However, this office relates to "the auditor of his Majesties Court of Wards and Liveries". John Tooke is known to have had at least two sons; one was Edward Tooke of the Inner Temple who bequeathed his estate to Mary (Halley) and her sisters. The other was Christopher Tooke who was Halley's father-in-law (*see* also MacPike).

The mystery of Mary Tooke's antecedents provides an interesting challenge to genealogists with time and patience. Among contemporary, but very shadowy, Tookes who may have had family connections with Halley's wife were the printer/publisher Tookes. Benjamin Tooke (1671-1723) was Jonathan Swift's publisher; his father, also a Benjamin (1642?-1716), started as a printer in Ireland and then became a book-seller in London from about 1669.

The elder Benjamin was publisher of Whiston's *New Theory of the Earth* (1696 edition) and served as treasurer to the private publishing monopoly of the Stationers' Company called the English Stock. In 1702 he was dismissed after an enquiry into irregularities but was fortunate to obtain the position of steward to St Bartholomew's Hospital. It is of interest that the senior "Ben Took" was often referred to in his lifetime as 'the Treasurer"; it is also known there was an elder sister called Mary. However, very little else is known about these particular Tookes.

Note 4 p. 48
While these discoveries about the three laws were a glorious triumph and demonstrated Kepler's genius, his other work, where he tried to build the universe round five Pythagorean solids and then the harmonies of the Pythagorean scale, which he set great store by, had no foundation of truth. Thus the minds of the great are sometimes as fallible as those of lesser beings.

Note 5 p.60
Like Halley, Edward Paget apparently was another of Flamsteed's "great disappointments". He was recommended by Newton to Flamsteed "as a very able mathematician". In turn Flamsteed befriended him and then recommended him for the post of Mathematical Master at Christ's school. At first he pursued his duties diligently, but, according to Flamsteed, within seven years he became a drunkard, lost all interest in mathematics and banished himself to India. One account claims he later returned to England to die from fever; another says he expired at Isphahan.

Note 6 p. 95
Halley and Gregory were not the only candidates for the Savilian Chair of Astronomy in 1691. Another was John Caswell (1656-1712) who was a graduate of Wadham College, Oxford. Like Halley he was unsuccessful when he applied for the Chair in 1691; he likewise was unsuccessful when he tried for the Mathematical Mastership of Christ's Hospital in 1695 (a post held by Edward Paget, *see* Note 5). However, when Gregory died in 1708, Caswell was appointed in his place. After their defeat in 1691, Halley was suspicious that Caswell, a friend of Flamsteed, bore him ill-will, but John Wallis told Halley to be open and friendly with him, for he did not share the Astronomer Royal's opinion.

Note 7 p. 96
Halley's first recorded disagreement with Flamsteed occurred *before* his St Helena exploits. In Hooke's *Diary* for 7 January 1675/6 we read: "... Flamsteed and Halley fallen out." Unfortunately Hooke tells us nothing about the nature of their disagreement, but their relationship must have been repaired, for later that year Flamsteed was lending assistance to Halley with his preparations for his journey to St Helena, and they were still relatively friendly to

each other until Halley returned from Danzig in 1679.

Hooke had a poor opinion of Flamsteed and on two separate occasions writes bluntly in his *Diary*: "Flamsteed mad."

Note 8 p. 110

Although little is known about Halley's friend Benjamin Middleton, he was by all accounts a seasoned voyager who had visited the West Indies and intimately knew those coasts made immortal by the Buccaneers. We can assume it was Middleton who kindled Halley's appetite for voyaging and geographical discovery. Reading between the lines in Hooke's laconic *Diary*, we can imagine those cozy evenings spent in Jonathan's Coffee House, in Change Alley, when Middleton and other seasoned voyagers were present and they spun out their yarns of past adventures to the Greshamites. Another sometime voyager present in this intimate circle was Sir John Narborough, whose accounts of the vagaries of the "magnetical needle" in the South Seas may have provided Halley's initial inspiration to mount a scientific voyage to study magnetism. Because an exact knowledge of magnetic variation and its change was considered a possible novel way to calculate a ship's longitude while at sea, Halley, still ambitious for fame, would recognize that here, like the southern stars, was yet another ripe scientific fruit waiting to be picked.

Note 9 p. 114

There is a curious parallel to Halley's first voyage in Dampier's own experiences aboard *Roebuck*. There was trouble with his crew and in particular with his first lieutenant, a man called George Fisher. When *Roebuck* reached the coast of Brazil, Dampier was so fed up with Fisher's behaviour he sent him ashore in irons where the Portuguese Governor, Don John Lancastrio, a descendant of some British Lancasters and sympathetic to Dampier's plight, incarcerated the offender. Fisher remained there for four months before being shipped back to England via Lisbon. On his return from the South Seas, Dampier had to face three courts-martial: one for what their Lordships considered harsh treatment of Fisher; a second for the alleged mistreatment of another crew member (a trumped-up charge); and the third for the loss of the *Roebuck* (an old, leaky vessel before starting out) which foundered off Ascension during the return voyage.

The court-martial's thirty-four captains, including Sir Cloudesley Shovell, were strongly prejudiced against Dampier and not prepared to accept the word of an ex-buccaneer against that of Fisher, a regular naval officer. Dampier's charge that Fisher had incited mutiny was dismissed, and the court fined Dampier the whole of his pay.

Note 10 p. 120

Very brief entries of Halley's activities (under two spellings of his name) can be found in the *Calendar of State Papers* Domestic Series Vol 1, 1702-3, 1916 (reprint 1968). We read on 26 October 1702: "Queen Anne command 200 L to

Mr Edmund Hawley who is going abroad on the Queen's special service." Then on 6 November 1702: "Edmund Halley; to Germany on private business with servants." (*See* also MacPike.)

Note 11 p.126
It is apparently not widely appreciated that David Gregory, Halley's rival for office in 1691, was also interested in the period of this comet. In part of a memorandum in his hand, dated March 1702/3 (before Halley published his *Synopsis*) he notes: "The Comet 1680/1 its Orb would be exacter or agree better with observation, if in stead of Parabolick it was Made or assumed Elliptick, so as to make the Comets period about 2000 years or between 1000 and 3000 years." Thus he was nearer the mark than Halley's period of 575 years.

Note 12 p. 133
While it was Maupertuis and Clairaut who ultimately provided the scientific evidence to convince Frenchmen and other Europeans of the validity of Newton's ideas, it was their sometime friend Voltaire who did much to popularize them on both sides of the Channel. After his visit to London (1726-9), Voltaire arrived back in Paris eager to denounce Cartesian "whirlwinds". Now his future ambition was to produce a book simplifying Newton's ideas so that even society women and café loungers might grasp them. When completed, his book (*Les Élémens de la Philosophie de Neuton*) was sent to Amsterdam where it was pirated. However, the publisher found it wanting and employed a Dutch mathematician to expand it. Soon after in 1738 it was translated into English and published in London as *Elements of Sir Isaac Newton's Philosophy* (opposite). Voltaire never did fully grasp Newton's theories. The story goes that, conscious of his shortcomings, he one day taxed Clairaut for his honest and candid opinion about his abilities as a man of science. Clairaut took him at his word. He replied that unfortunately his friend Voltaire really had no gift for the subject and believed no matter how hard he continued to study it, he would not improve much. After this Voltaire rationalized his fate and realized his true bent "was the human race, past, present and to come". (*See* also *Note* 17.)

Note 13 p. 133
Many generations after Fontenelle's death, animosity towards Newton might still be conjured up among nationalistic Frenchmen. When, circa the 1860s, Vrain-Denis Lucas, the prince of forgers, concocted a whole series of letters purporting to demonstrate that Blaise Pascal had discovered the laws of gravitation as early as 1652 (thus predating Newton's discoveries by more than thirty years), a certain gullible geometrician, Michel Chasles (1793-1880), was readily duped...

In due course, Chasles announced to the Academy that he was going to publish a book to support this claim. After examining the letters, other Academicians were also very willing to believe it, for who among them could

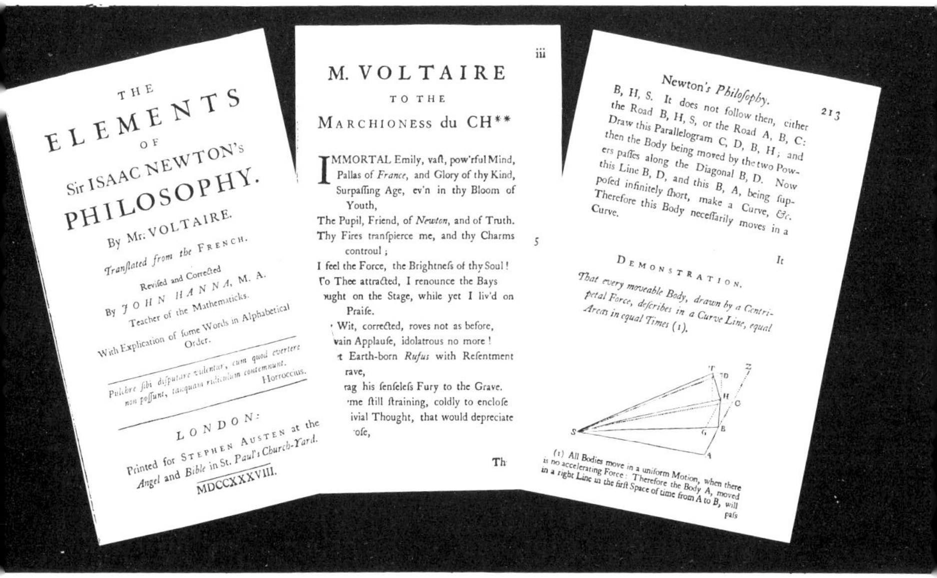

THE
ELEMENTS
OF
Sir ISAAC NEWTON'S
PHILOSOPHY.
By Mr. VOLTAIRE.
Tranſlated from the FRENCH.
Reviſed and Corrected
By JOHN HANNA, M. A.
Teacher of the Mathematicks.
With Explication of ſome Words in Alphabetical Order.

Pulchre ſibi diſputare videntur, cum quod evertere non poſſunt, tanquam ridiculum contemnunt. Horroccius.

LONDON:
Printed for STEPHEN AUSTEN at the *Angel* and *Bible* in St. *Paul's Church-Yard.*
MDCCXXXVIII.

iii

M. VOLTAIRE
TO THE
MARCHIONESS du CH**

IMMORTAL Emily, vaſt, pow'rful Mind,
Pallas of *France*, and Glory of thy Kind,
Surpaſſing Age, ev'n in thy Bloom of Youth,
The Pupil, Friend, of *Newton*, and of Truth.
Thy Fires tranſpierce me, and thy Charms controul ;
I feel the Force, the Brightneſs of thy Soul !
To Thee attracted, I renounce the Bays
...ught on the Stage, while yet I liv'd on Praiſe.
... Wit, corrected, roves not as before,
...vain Applauſe, idolatrous no more !
...t Earth-born *Rufus* with Reſentment ...rave,
...rag his ſenſeleſs Fury to the Grave.
...me ſtill ſtraining, coldly to encloſe
...ivial Thought, that would depreciate ...oſe,

Th

Newton's *Philoſophy.* 213

B, H, S. It does not follow then, either the Road B, H, S, or the Road A, B, C: Draw this Parallelogram C, D, B, H; and then the Body being moved by the two Powers paſſes along the Diagonal B, D. Now this Line B, D, and this B, A, being ſuppoſed infinitely ſhort, make a Curve, &c. Therefore this Body neceſſarily moves in a Curve.

It

DEMONSTRATION.

That every moveable Body, drawn by a Centripetal Force, deſcribes in a Curve Line, equal Areas in equal Times (1).

(1) All Bodies move in a uniform Motion, when there is no accelerating Force: Therefore the Body A, moved in a right Line in the firſt Space of time from A to B, will paſs

150 Voltaire was instrumental in popularizing Newton's ideas in France when he wrote his *Les Élémens de la Philosophie de Neuton.* (*left*) It was published in Amsterdam where it was pirated and revised. (*below*) In 1738 it was translated into English and appeared in London as *Elements of Sir Isaac Newton's Philosophy.* Note the dedication to Voltaire's blue-stocking mistress the Marquise du Châtelet. The right folio in the English translation is open at Newton's classical demonstration of centripetal force and Kepler's law of equal areas (*see* also notes 12 and 17).

ELÉMENS
DE LA
PHILOSOPHIE
DE NEUTON,
Mis à la portée de tout le monde.
Par Mr. DE VOLTAIRE.

A AMSTERDAM,
Chez ETIENNE LEDET & Compagnie.
M. DCC. XXXVIII.

a)

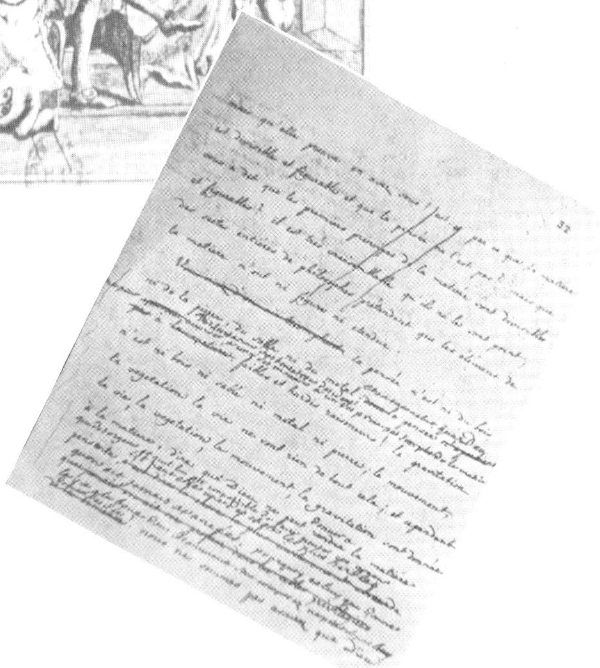

b)

deny a Frenchman the glory that so far had been all Newton's.

The Pascal claim was only part of Lucas's audacious fraud which, by comparison, dwarfed into insignificance that of his latter-day counterparts. In his collection were 175 letters from Pascal to Newton, 76 to Boyle and 130 to Galileo; many hundreds of others were also passed off as true holographs of notables of the past, including Roman emperors and Greek philosophers. Surprisingly, at first no one in the Academy queried why they were all written in French!

A leading figure in the final exposure was Le Verrier, who by diligent scholastic detective work tracked down some of the textual sources which had been copied verbatim by Lucas to give his forgeries verisimilitude.

Note 14 p. 135

It is of interest that Jonathan Swift in his *Gulliver's Travels* (1726) had the astronomers of Laputa know the periods of 93 comets. This was, of course, written thirty-two years before Halley's prediction was proved right and suggests in this instance a satirical dig at Halley (*see* also *Notes* 3 and 20).

Note 15 p. 140

Johann "George" Palitzsch (1723-88) is sometimes portrayed as an ignorant rustic with little learning, but the opposite is the case. He was a gentleman by birth and had many diverse scientific interests, including an expertise in botany and lightning conductors. His library is said to have consisted of 3,500 volumes. While the Seven Years War raged all round Dresden, he buried his telescopes for safe keeping and only disinterred his "seven-foot" instrument in order to search for Halley's comet. After he rediscovered the comet in 1758, he became a celebrated figure and corresponded with the leading societies in Europe and individually with many astronomers, including William Herschel. He is remembered today by a monument in Prohlis.

Note 16 p. 140

Delisle must have had his reasons for suppressing Messier's pre-perihelion observations of Halley's comet in 1759. However, what these were remain one of those intriguing astronomical mysteries inherited from the past (leastwise so far in English-language comet literature). It is known that Delisle was a visitor to England in 1724 when he met Newton and Halley, and he was apparently soon on very good terms with both of them. Newton was much impressed with his learning; he proposed him for election to the Royal Society and then presented him with a portrait. It is also known Delisle carried out a voluminous correspondence with other astronomers, much of which still exists in manuscript in Paris. Hopefully, in the future, some French archivist may be able to shed light on Delisle's strange motivation.

151 (*a*) A romanticized old French print showing Voltaire totally absorbed in his task when visited by the Muse Urania during the writing of his book to popularize Newton's ideas. (*b*) A folio from Voltaire's original manuscript.

Note 17 p. 142
Who this personage was is uncertain, for in their time both Clairaut and Lalande inspired numerous "bluestockings"—a class of superior female practically unknown on the other side of the Channel round this period. Perhaps Clairaut's most noteworthy pupil was Gabrielle Marquise du Châtelet-Lomont (1706-49), who became Voltaire's mistress in 1733. She studied mathematics with Maupertuis and then with Clairaut before embarking on her translation of Newton's *Principia.* She was likely Voltaire's superior in mathematics. He dedicated his own *Elements of Sir Isaac Newton's Philosophy* (1738) to her, and it is probable it was she who helped him properly understand and then render into common language, Newton's main ideas (*see* Figs 150 and 151).

Note 18 p. 142
In later years Lalande wrote at some length about the strong camaraderie that developed between him and Nicole Lepaute, recounting: "She drew me away from dangerous alliances; she brought me into a circle of pleasant well-educated people; she suffered my faults and helped me overcome them. The times I spent with her and her family are those I like best to recall, and this memory spreads a gentleness over the last years of my life as her friendship gave charm to my youth."

Lalande is a colourful figure in the history of astronomy. Apart from his supposed predilection for eating spiders, he missed a greater claim to fame by his failure to recognize Neptune fifty-one years before it was identified in Berlin in 1846. According to a manuscript at the Paris Observatory, Lalande plotted a star of 8^m on 8 May 1795 while observing stars for his *Histoire Celeste.* Two days later he re-examined the same area of sky but not finding this earlier star fixed in the same place and noting another not far away, he doubted his first observation. He decided to strike out the first star and to retain the second, but he even marked this one as doubtful. After Neptune was found and its orbit traced back, it was discovered that Lalande's now missing stars had been the planet.

Note 19 p. 144
Another anecdote about Messier relates how at night he supposedly never took his eyes from the sky and as a result once fell down a well (or ice-house?) in the president's gardens. The accident laid him up for months while all the time he grieved for the comets his disability caused him to miss. He developed a close friendship with Lalande, and it was from him, it is said, Messier would often borrow money or lamp oil when his small stipend ran out or was overdue.

152 (*a*) Voltaire (1694–1778). (*b*) Gabrielle Émilie Marquise du Châtelet-Lomont (1706–49), Voltaire's blue-stocking mistress who later translated Newton's *Principia* into French with the help of Maupertuis and Clairaut.

a)

b)

Note 20 p. 150
The idea of cometary impact with the Earth has exercised the human imagination in all ages into the present and provides a rich source for pseudo-scientific speculation. In Jonathan Swift's time the subject of cometary catastrophe (due to Whiston and indirectly to Halley) was discussed and feared by the common man as nuclear war is today. In his *Gulliver* satire, Swift tells us how the Laputans were in perpetual alarm over the comet danger and could neither sleep quietly nor did they have relish for the common pleasures or amusements of life. A famous Victorian-age comet scare occurred in 1857 when it was predicted that a great comet seen in 1264, which had a tail 100° in length, would make its return. Much publicity was engendered among the lay public. A German astrologer predicted it would collide with the Earth, and an exact date of 13 June was fixed by a Belgian almanac-maker. The anticipated encounter gave rise to a plethora of scare-mongering literature, and a typical example was a sixpenny pamphlet entitled *Will The Great Comet now Rapidly Approaching Strike The Earth?* (Fig 62). However, to the relief of the ignorant, the comet never materialized.

The effects of a comet impact may in truth have wide-reaching influences in shaping the destiny of Planet Earth. All manner of events in the geological/biological record have been attributed to such cataclysms occurring in prehistoric times. The demise of the dinosaurs 60 million years ago and other significant biological extinctions are now fashionably attributed to direct cometary impacts or brushes by comets with our atmosphere. Likewise the origin of life may be comet-borne to the Earth (*see* Chapter 18). Nevertheless, all such theories are at present conjectural, and there is little hard evidence to support them.

The Tunguska "meteorite" which fell on Siberia in 1908 is now widely believed to have been caused by the Earth encountering a small fragment of a comet, possibly Encke's (*see* Note 24).

Note 21 p. 153
The closest known approach of Halley's comet to the Earth, which occurred at its AD 837 apparition, was about ten times the Moon's distance.

Note 22 p. 157
While Messier had earned the title "the Ferret", Pons was soon nicknamed "the Comet Magnet". By a curious coincidence the last comet discovered by Messier (the August comet of 1801) was also the first independently discovered by Pons. What makes Pons a unique figure is the fact that until he was twenty-seven years old he had no astronomical interests. It was when he became door-keeper to the Marseilles Observatory his latent interests were kindled, and soon after he was taking lessons from Thulis, the Director. At the age of forty he discovered his first comet, and from then on, until 1827, it was a rare year that failed to register the discovery of at least one comet by him. During the eight-month period February to August 1808 he found no less than

153 A nineteenth-century artist's dramatic rendering of the returning comet of 1264 predicted to collide with the Earth on 13 June 1857.

five comets and another five in the twelve-month period August 1826 to August 1827.

Note 23 p. 159
For his outstanding work Encke was awarded the Gold Medal of the Royal Astronomical Society in 1824, and the comet thereafter became known as Encke's comet. However, history relates that Encke himself always more modestly referred to it as "the Comet of Pons". In comet catalogues today it is designated P/Encke, the P prefix denoting it is periodical and a comet whose orbit is known with certainty. In the Soviet Union the comet is referred to as P/Encke-Backlund to commemorate the study made of its movements between 1819 and 1891 by the astronomer Oskar Backlund of the Pulkova Observatory. Backlund received the Gold Medal from the Royal Astronomical Society in 1909.

Since Encke's time the comet has been observed at every apparition except one during its short $3\frac{1}{3}$-year journey round the Sun. It was only missed at its apparition of 1944 during World War II when it was unfavourably placed for observation in the northern hemisphere.

Note 24 p. 160

When at some unknown, prehistoric period Encke's comet was captured from a long-period orbit (*see* Fig 65) into its smaller present-day $3\frac{1}{3}$-year orbit, it was probably a much larger comet than it is today. Since its capture it has probably suffered several major disruptions. Part of the resulting debris is seen annually in the display of the Beta Taurid meteors occurring about 30 June each year. It has been suggested that one of the more sizable chunks which broke off the comet, and then later encountered the Earth, was the mysterious object that caused the famous Tunguska event in 1908 (Fig 155). Another interesting suggestion links Encke's comet with the asteroid 2212 Hephaistos as both objects move in very similar orbits.

154 The probable track of the Tunguska object.

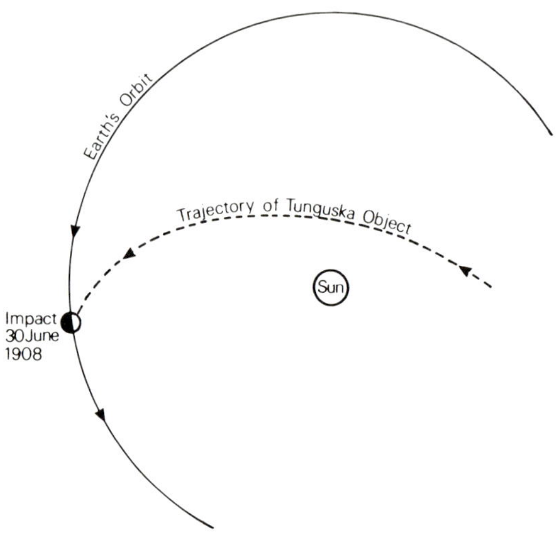

Note 25 p. 163

Friedrich (Franz) Wilhelm Bessel (1784-1846) was one of the giants in German astronomy. He had a lifelong interest in Halley's comet. His first paper (published in von Zach's ***Monatliche Correspondenz*** in 1804) was a computation of Harriot's observations of Halley's comet in 1607. One of his other consuming interests was in the supposed planet which he and many others believed lay beyond Uranus. In 1840 he was deeply immersed in computing its orbit. The following year his only son, an astronomer of great promise, died,

155 Tunguska, Siberia, the site of a strange cosmic-inspired cataclysmic event which occurred on 30 June 1908 and was probably caused by a cometary impact. (*inset*) Note the trees felled by the tremendous air-blast.

and Bessel lost heart in his work. However, a few years later, after visiting England to attend a British Association meeting and being spurred on by his friend John Herschel, he began to refine his calculations, helped by his young assistant Flemming. He was now on the verge of success, but unfortunately Flemming too died suddenly and then Bessel shortly after on 17 March 1846. Neptune was discovered by Galle and d'Arrest in Berlin on 23 September 1846 as a result of the computations by the brilliant French mathematician Le Verrier (1811-77).

Note 26 p. 209
As may be expected, newspapers at the time provided some colourful anecdotes about the event. One US newspaper recounted that the adverse publicity about the discovery of cyanogen in Morehouse's comet two years previous (*see* p 213) alerted several ultra-cautious New Yorkers to the possibility of asphyxiation because the gas in the tail of Halley's comet might pollute the Earth's atmosphere. To sustain life during the fatal night, when the Earth would pass through the tail of the comet, a run began on supplies of bottled oxygen. According to the same newspaper, supplies proved inadequate, and some neurotics, unable to find the life-sustaining gas, committed suicide.

Note 27 p. 211
The so-called Sungrazer comets form one of the most remarkable groups of comets ever studied. They derive their name from their characteristic close perihelion approach to the Sun's surface—actually skimming through its outer tenuous atmosphere. All have similar orbits which lie in an approximate direction towards the bright star Sirius. Their periods of revolution are measured in some thousands of years. They all probably originate from one gigantic parent comet which in the past divided and then successively subdivided at each return to perihelion. Such subdivisions have been seen with several Sungrazers near perihelion. The Sungrazer 1882 II subdivided and gave birth to at least eight satellite comets, and in more recent times the Ikeya-Seki Sungrazer 1965 VIII split into two components. Perhaps one of the most remarkable of the Sungrazers seen in historical times was that seen in 1843 which developed an enormous tail stretching over 70° (Fig 89); this in terms of its true length was equivalent to the distance from Mars to the Sun.

The super-brilliant comets reported in Greco-Roman times were probably all Sungrazers. One prodigious comet seen in 371 BC, mentioned by Diodorus Siculus, broke into two parts. But the brightest comet ever was that seen in the reign of Mithridates (132-63 BC), described by contemporaries as "brighter than the Sun". It was supposedly followed for seventy nights and was so luminous it blotted out a quarter of the visible stars. Although the exact year of its appearance is uncertain, it was undoubtedly a real comet and is portrayed in a copper coin of the period. When one day its highly eccentric orbit returns it to the Sun, it will provide onlookers with the unique opportunity of viewing the most spectacular object ever to visit our night skies.

156 Another view of the Sungrazing comet of 1843 as seen from Blackheath in south London. In terms of its true length the tail stretched equal to a distance from the Sun to the orbit of Mars.

The Sungrazers may provide one of the few instances of the ancient Egyptians' observations of comets which are not cited directly in their literature. It is likely the legends of the Egyptian Phoenix or Bennu bird (seen in hieroglyphs) are direct allegories to this brilliant group of comets. Old legends relate that the bird's plumage, just before its destruction, was of a beautiful golden colour—as is exactly the case with Sungrazers (due to sodium emissions) when seen from Earth just before they are apparently swallowed up by the Sun and then again after they re-emerge round the opposite limb.

It is of interest that the early Christians also cited the Phoenix as the emblem of destruction and resurrection.

Note 28 p. 238

One must stress again that all computed apparitions for Halley's comet pre-240 BC must be open to some doubt, and in studying them one is constantly reminded of the glib Edinburgh correspondent circa 1835 (*see* p 178). When Crommelin became interested in Halley's comet in the early 1900s, he was surprised at the wide discrepancy between Pontécoulant's perihelion prediction for 1910 and that of Ångström's for 1912-13. This was part of Commelin's own motivation for taking up a definitive study of the comet's orbit, for, before embarking on the task, he aptly remarked: "One can hardly imagine a greater loss of prestige to astronomy that that which would arise if there were a notable error in the prediction of this return of the comet, after the wonderful success achieved in 1759 and 1835..."

Crommelin and the others in his team generally vindicated Pontécoulant's calculation (24.37 May, Paris civil time—though this precise figure is still open to dispute due to uncorrected nineteenth-century printer's errors), it being a little over one month adrift (as he had calculated the perturbations only approximately). Ångström's method involved using Hind's perihelion dates, linking them all from 12 BC to AD 1835 so as to derive a mean period and construct a curve to fit. While this fitted the previous 25 successive passages admirably, it broke down completely when it was extended to predict the new, twentieth-century return, introducing an error of 2-3 years!

Note 29 p. 242

Cowell's and Crommelin's efforts marked the closing phase of the golden age of orbit computing which had begun in the seventeenth century by their countrymen Newton and Halley. By all accounts Crommelin, the dominant influence, was often tetchy and unapproachable when he wrestled with a tricky problem in celestial mechanics.

One prime font of Crommelin memorabilia was L J Comrie (1893-1950), in his time a brilliant but irascible mathematician who blotted his copy-book while in high government service and was forced to resign his post but then put his misfortune to good use and formed the world's first private scientific computing company. As Crommelin's one-time colleague, he knew him well and in due course, after Crommelin died in 1939, inherited some of his key mathematical papers and a fund of good stories. Comrie once related to me how Crommelin had never come to terms with the "3-day error" incurred in the prediction for Halley's comet in 1910. Like Pontécoulant before him, he was keen to have a shot at getting the next apparition right.

According to Comrie, Crommelin—using another precedent, this time Rosenberger's in 1835—finally settled on two results. If he relied simply on Newtonian/Keplerian equations of motion, then 6 February 1986 was the date indicated. However, according to Comrie, he really preferred the alternative date that made use of what he termed the "2- to 3-day anomaly factor". When this was applied, it would bring the comet back sometime on 9 February 1986... *Requiescat in pace!*

Note 30 p. 279

Popular journalism, as a whole, has always been a mischievous influence in keeping the public at large informed about comets. All too often lay journalists, in their ignorance, confuse rather than clarify. When in January 1910 the Great Daylight comet suddenly appeared at the same time as Halley's comet was gracing the sky, a celestial comedy of errors was enacted at newspaper level.

Previously in the popular press all talk had been centred round Halley's comet. Then fate intervened, and a garbled agency message from South Africa hummed over the wires with the story of the discovery of "Drake's comet" (instead of a "great comet"). Who, newsman queried, was this mysterious Drake? Halley, yes, they vaguely knew about him as the Englishman who had predicted the return of that comet now fast approaching the Sun...but Drake? Yet on second thoughts there was *a* Drake, an historical figure really much better known than Halley, for had not Sir Francis Drake helped destroy the Armada! However, not even the most resourceful newsman could find any record of Sir Francis ever having discovered a comet.

Through late January 1910, newsmen, still confused, could not quite agree among themselves what to call this bright apparition out of the blue, but "Drake's comet" held place as firm favourite. By now some actually queried whether perhaps even the astronomers themselves were confused. In the end it became the "Daylight comet", or the "Miners' comet", but as it quickly faded, its advent turned into a nine-day press wonder.

Note 31 p. 282

Genuine colour photographs of Halley's comet did not exist before its mid-1980s apparition. Although colour films had been invented in the early years of the twentieth century, their speed was still too slow to be used to photograph the comet in 1910.

While colour-sensitive photographic plates provide dramatic renderings of comets, these are not a true measure of genuine colour as seen through a telescope or with the naked eye. However, on rare occasions a bright comet near the Sun may have a genuine yellowish tinge owing to sodium emissions triggered off in the head (*see* note 27), or more rarely a comet may take on a greenish hue owing to magnesium emissions. Halley's comet at perihelion never approaches close enough to the Sun for demonstrably true colour effects to be noted, and even at apparitions where it develops a long tail it appears in a mundane greyish-white garb.

In ancient European records and in Chinese annals there are several accounts of coloured comets. In 49 instances where colours are recorded in one Chinese annal, 23 were described as white, 20 bluish, 4 red or reddish-yellow and 2 greenish. According to Seneca the comet of 146 BC (which, of course, he did not see himself) was "of a fiery red" and "as large as the Sun". Pliny frequently writes of comets whose "mane" [tail] was the colour of blood. According to Pingré, during the Middle Ages there were several blue or black comets.

How truthful then are these older observations? The problem is complicated because many comets of earlier times were actually observations of auroral displays, and these occasions give rise to practically every colour in the spectrum. Meteorological effects may also influence the colour when a comet is low down near the horizon. A red sunset may impart a definite ruddy tint, sometimes even green, to a bright comet. Volcanic dust suspended in the

Earth's atmosphere also had this effect on colour, and "blue" moons are not an uncommon report at these times.

Although one day a genuinely blue or red comet may come along to astound us, on present available evidence it seems that those vivid colour descriptions of the past were little more than an ancient author's feverish device to dramatize a particular apparition. They probably have as little verity as those lurid colour depictions seen only in books and on TV during 1985-6...

Acknowledgements and Credits:
The author and publisher gratefully acknowledge and credit the following figure sources. Aries Press Archives, 2–22, 25–38, 40–64, 66–9, 71–86, 88–90, 92–6, 98, 102–110, 118, 120, 123–5, 128, 130, 136, 150–3, 155–6; *Time*, 136; Goddard Space Flight Centre, 117, 122; United Features Syndication, 127; Australian Information Service, 116, 144, 146; Gerald Hawkins, 145; *East Anglian Daily Times*, 126 (lower); Cerro Tololo Inter-American Observatory, 113, 115; Boyden Observatory, 112; Information Technology Limited, 138; Tokyo Observatory, 114; British Aerospace, 133, 139; Novosti Press Agency, 129, 134, 147–8; Royal Astronomical Society, 99–101; Akira Fujii, 140; Mt Palomar Observatories, 121, 149; European Space Agency, 141–2; *Daily Mail*, 126 (upper); *Manchester Guardian*, 137, 143; Author, 22–4, 39, 70, 87, 91, 97, 111, 154, 131–2.

The author is especially grateful to his long-suffering wife, Johanne, for her invaluable suggestions and her generous and unflagging editorial assistance.

Photoset in 12 on 13 point Garamond Scantext by Hugh Wilson Typesetting, Norwich, NR2 1LT.
Printed in Great Britain by The Garden City Press Limited, Letchworth, Hertfordshire SG6 1JS.

APPENDIX 1

Orbital Geometry of a Comet

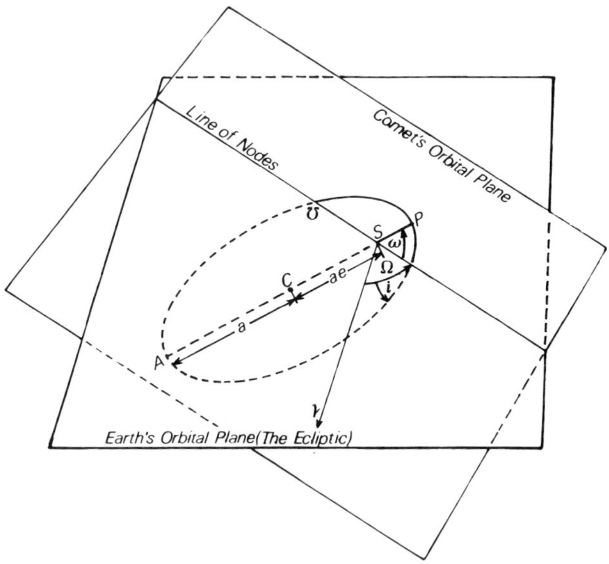

The Orbit of a Comet.

S = Sun.

γ = First point in Aries (the reference-marker to fix the orbit in space).

P = Perihelion.

A = Aphelion

a = half the major axis of the ellipse.

ae = eccentricity.

ω = argument of perihelion (the orientation from the ascending node in relation to the perihelion).

☊ = ascending node longitude.

☋ = descending node.

i = inclination of the comet orbit to the ecliptic.

To define a comet's path round the Sun requires knowledge of six quantities or elements.

1) The perihelion distance of the comet (or distance from the Sun at its closest approach).
2) Its heliocentric longitude (a position in relation to the Sun usually termed the argument of perihelion, *see* Fig *above*).
3) The position of the nodes (or where the path of the comet crosses the plane of the ecliptic).
4) The inclination of the comet's orbit to the plane of the ecliptic.
5) The exact time of perihelion passage.
6) The period of the comet's revolution.

See also Fig *above*.

Unlike the major planets, many comets move in paths that are inclined to the ecliptic at steep angles. When their inclinations exceed 90°, comets are said to possess retrograde orbits round the Sun (e.g. Halley's comet which is inclined at 162°.2). These retrograde comets revolve in clockwise direction, i.e. the opposite (reverse) motion to that of the planets.

The distinction between short- and long-period comets is hazy, but one may say any comet with a period of *less* than 150-200 years is one of short period. Some long-period comets which have highly elongated elliptical paths, such as the Sungrazers, can never be given accurate orbital periods owing to the close similarity of their ellipses to a true parabola; thus their orbital periods round the Sun may range between anything from 1,000 to 100,000 years or more.

APPENDIX 2

Comet Designations

Comets discovered in any one year are provisionally tagged, in order of discovery, by the letters *a*, *b*, *c*, etc together with the name of the discoverer, viz comet Humason 1961*e*. Thus Humason's comet was the *fifth* comet discovered in the year 1961. At a later date, when full orbital details are known, the comet is renumbered in the strict chronological order it came to perihelion in that year, or in a later year, and affixed with a Roman numeral I, II, III, etc. Thus comet Humason is entered in the permanent catalogues: Comet 1962 VIII Humason 1961*e*, as the eighth comet in 1962 to come to perihelion (which one immediately sees it did not do in the year of its discovery). If a comet is independently discovered by a number of people, it is designated by the names of the first three observers to report it, e.g. Tomita-Gerber-Honda 1964*c* (VI).

Periodic comets can be recognized in catalogue nomenclature and in literature by the prefix P; thus comet 1910 II is fully listed as: P/Halley 1910 II (1909*c*). Note that although Halley's was the third comet resighted in 1909, it did not come to perihelion until 1910. When Halley's comet was resighted in 1982, it was provisionally tagged 1982*i*, but its final designation for its 1986 apparition will not be known for a year or so after its perihelion passage in February 1986. Note that a name tag is only applied to discoveries of new comets, not returning periodic comets which (with few exceptions) are still identified by the name of their original discoverer(s).

A comet may also bear a name other than that of its original discoverer(s), and Halley's comet provides the most famous example. Likewise, Encke's comet was first discovered by the Frenchman Méchain (in 1786).

Select Sources and References

The prime, but unfortunately far from complete, modern source for Halleiana is contained in E F MacPike, *Correspondence and Papers of Edmond Halley*, (Oxford 1932; London 1937). This interesting work provides an extensive list of primary source-material and is the basis for all modern scholarship.

Extremely useful secondary sources which contain general accounts of Halley's various scientific achievements are Angus Armitage, *Edmond Halley*, London 1966; and Colin Ronan, *Edmond Halley: Genius in Eclipse*, London 1969. An excellent account of Halley's voyages, containing references not cited elsewhere, is in *The Three Voyages of Edmond Halley in Paramore 1698-1701* 2 VOLS, edited by Norman J W Thrower and published by the Hakluyt Society (2nd Series 156-7), London 1981.

In spite of the above, the primary and secondary source-material for Edmond Halley's life is severely limited, and there exist lacunas spanning several years at a time. An article on Halley contained in *Biographia Britannica* (1757) is thought to have been written by Thomas Birch or John Machin (one-time Secretary to the Royal Society) from material prepared by Halley's son-in-law, Henry Price, who was married to Catherine Halley (1688?-1765).

In the nineteenth century a bundle of manuscripts relating to Halley's life contained in material left by John Machin was lodged in the Library of the Royal Astronomical Society in London, but this subsequently disappeared and now, unfortunately, cannot be traced.*

Among the numerous biographies of Newton, two of the more modern and readily accessible to general readers are Richard S Westfall, *Never at Rest*, Cambridge 1980; and Frank E Manuel, *A Portrait of Isaac Newton*, London 1980. The indispensable source for all serious students of Newton (with limited use also for Halley and Flamsteed) is supplied by the monumental *Correspondence of Isaac Newton* 7 VOLS, Cambridge 1975. These collected letters and memoranda (with translations into English) also offer an excellent source of browsing for interested general readers who already have some familiarity with the scientific achievements of the period.

Note that the above primary and secondary sources provide only brief general references to Halley's comet.

*For a comprehensive account of this intriguing mystery *see* Peter Lancaster Brown, *The Comet Man*: The *Life* and *Strange* and *Surprising Adventures* of *Edmond Halley*, Aldeburgh 1987

Other Select Sources and References

Anglo-Saxon Chronicles (1066)
Armitage, Angus, *John Kepler*, London 1966
Baily, Francis, *John Flamsteed*, London 1835
Bainbridge, John, *An Astronomical Description of the late Comet*, 1619
Bayle, Pierre, *Lettre sur la Comète*, 1682
Pensées diverses sur la Comète, 1683
Berkeley, George, *The Analyst* (an attack on the infidel mathematician Edmond Halley), London 1734
Brady, Joseph, *Journal of the British Astronomical Association*, VOL 92
Calder, Nigel, *The Comet is Coming!*, London 1980
Central Bureau for Astronomical Telegrams, Cambridge, Mass, USA (Director: Brian Marsden)
Chambers, George F, *The Story of the Comets*, Oxford 1909, 1910
Clairaut, Alexis-Claude, *Théorie des Comètes*, Paris 1760
Comets, Asteroids, Meteorites: interrelations, evolution and origins, IAU Colloquium No 39, 1976, edited by A H Delsemme, Toledo 1977
Cowell, P H and Crommelin, A C D, *Essay on the Return of Halley's Comet*, Leipzig 1910
Dick, Oliver Lawson, *Aubrey's Brief Lives*, London 1960
Evelyn, John, *Evelyn's Diary*, Oxford 1955
Gadbury, John, *De Cometis*, London 1665
London's Deliverance Predicted, London 1715
Guillemin, Amédée, *The World of Comets*, London 1877
Hearne, Thomas, *Remarks and Collections*, IV, Oxford 1884-1921
Hevelius, Johannes, *Cometographia*, Danzig 1668
Annus Climactericus, Danzig 1685
Hind, J Russell, *On the Expected Return of the Great Comet of 1264 and 1556*, London 1848
Monthly Notices of the Royal Astronomical Society, 1850
The Comets, London 1852
History of Science, VOL 3, Cambridge 1964
Hooke, Robert, *The Diary of, 1672-1680*, London 1968
Hoyle, F and Wickramasinghe, N C, *Lifecloud*, London 1978
Diseases from Space, London 1979
Evolution from Space, London 1981

Space Travellers—the Bringers of Life, Cardiff 1981
Hoyle, F, *The Intelligent Universe*, London 1983
Icarus, VOL 23, Number 4
Kepler, Johannes, *De cometis*, 1619
Kiang, T, *Observatory*, VOLS 91, 93
Koestler, Arthur, *The Sleepwalkers*, London 1959
Kuiper, G P and Roemer, E, *Comets: Scientific Data and Missions*, Proceedings of the Tucson Comet Conference, Tucson 1972
Lalande, J J, *Bibliographie Astronomique*, Paris 1803 (reprinted 1970)
Lancaster Brown, P, *Journal of the British Astronomical Association*, VOLS 75, 76
Comets, Meteorites & Men, London 1973, New York 1974
Megaliths, Myths & Men—an introduction to astro-archaeology, Poole, New York 1976
Megaliths & Masterminds, London, New York 1979
Halley & his Comet, Poole, New York 1985
The Comet Man: The *Life* ... of *Edmond Halley*, Aldeburgh 1987
Lilly, William, *History of his Life and Times*, London 1715
Lindsay, Jack, *Origins of Astrology*, London 1971
Lubienieczki, Stanislas, *Theatrum Cometicum*, Amsterdam 1667 and 1668
Lynn, W T, *The Observatory*, VOL 14, 347-8
Lyttleton, R A, *The Comets and their Origins*, 1953
MacPike, E F, *Hevelius, Flamsteed and Halley*, London 1937
Marsden, Brian, *Astronomical Journal*, 72, No 9, 1967
Science, VOL 155, pp 1207-1213, 1967
Maupertuis, P L Moreau de, *Lettre sur La Comète*, Paris 1742
Milne, David, *Essay on Comets*, Edinburgh and London 1828
NASA, *The Study of Comets*: A conference held at Goddard Space Flight Center Greenbelt, Maryland 28 October-1 November 1974, Washington 1976
Nature, VOL 300, 239, 1982
Needham, Joseph, *Science and Civilization in China*, VOL 3, Cambridge 1959
Newton, Sir Isaac, *The Mathematical Principles of Natural Philosophy*, translated by Andrew Motte, London 1729
Newton, Sir Isaac(?), *A Dissertation on Comets*, London *c* 1750
Pepys, Samuel, *The Diary of Samuel Pepys*, (edited by Robert Latham and William Matthews), London 1983
Plummer, H C, *Nature*, VOL 150, 249
Porter, J G, *Comets and Meteor Streams*, London 1952
Proctor, Richard A, *St Paul's Magazine*, September 1871
Rawlinson, George, *History of Herodotus*, London 1880
Richter, N B, *The Nature of Comets*, London 1963
Secretan, the Rev C F, *Memoirs of the Life and Times of the pious Robert Nelson*, 1860

Seneca, *Physical Science in the Time of Nero* (*Quaestiones Naturales*), translated by John Clarke, London 1910
Sky and Telescope, VOL 64, 551, 1982
Tacke, Johann, *Coeli Anomalon, id est, De Cometis*, 1653
Turner, H H, *Halley's Comet*, Oxford 1908
Vsekhsvyatskii, S K, *Physical Characteristics of Comets*, Jerusalem 1964 (although not always accurate or complete, a prime source-book for past comet apparitions)
Whiston, William, *The Cause of the Deluge Demonstrated*, London 1714
Memoirs of the Life and Writings of Mr William Whiston, London 1749
Yeomans, Donald K, *The Comet Halley Handbook*, Washington 1982
See also text, passim

Index

Figures in italics refer to page numbers of selected illustrations.